London Mathematical Society Lecture Note Series. 12

PROCEEDINGS OF THE SYMPOSIUM ON
Complex Analysis
CANTERBURY 1973

Dov Aharonov and Harold S. Shapiro, Lars V. Ahlfors, Albert Baernstein II, D. A. Brannan, Leon Brown and Allen Shields, David Drasin, Peter L. Duren, Albert Edrei, Matts Essén, Matts Essén and Daniel F. Shea, W. H. J. Fuchs, F. W. Gehring, H. Grunsky and James A. Jenkins, Simon Hellerstein and Daniel F. Shea, A. Huber, James A. Jenkins, J. P. Kahane, J. Korevaar, Ü. Kuran, A. J. Lohwater and G. Piranian, Zeev Nehari, D. J. Newman, Ch. Pommerenke, L. A. Rubel, T. B. Sheil-Small, Uri Srebrö, Paul Turan, Allen Weitsman, L. C. Young, Lawrence Zalcman, W. K. Hayman.

Edited by J.Clunie and W.K.Hayman

Cambridge · At the University Press · 1974

CAMBRIDGE UNIVERSITY PRESS
Cambridge, New York, Melbourne, Madrid, Cape Town, Singapore, São Paulo

Cambridge University Press
The Edinburgh Building, Cambridge CB2 8RU, UK

Published in the United States of America by Cambridge University Press, New York

www.cambridge.org
Information on this title: www.cambridge.org/9780521204521

First published 1974
Re-issued in this digitally printed version 2008

A catalogue record for this publication is available from the British Library

ISBN 978-0-521-20452-1 paperback

Contents

Preface

This is the proceedings of a symposium in complex analysis held at the University of Kent, Canterbury, 1973. The financial support for this symposium came from S. R. C., NATO and the University of Kent. The organising committee consisted of J. Clunie, W. H. J. Fuchs, W. K. Hayman, M. E. Noble, Ch. Pommerenke and J. Shackell.

The first part of the proceedings contains contributions from the participants in the symposium. Almost all of these contributions are abstracts of talks given at the symposium.

A similar conference was held at Imperial College in 1964 and arising from this W. K. Hayman collected a number of problems and these were published by the Athlone Press in 1967 with the title 'Research problems in function theory'. In the second part of the proceedings W. K. Hayman gives a progress report on work on these earlier problems, and then gives a list of problems that arose from the present symposium.

The organisers of the symposium would like to thank Miss Christine von Baumann at the University of Kent and Miss Pat Edge at Imperial College for the secretarial help they have provided. Such help was given before and during the symposium and also with the preparation of these proceedings.

J. C. and W. K. H.

A MINIMAL-AREA PROBLEM IN CONFORMAL MAPPING (ABSTRACT)

DOV AHARONOV and HAROLD S. SHAPIRO

Let U denote the open unit disc of the complex z-plane. For a function f holomorphic in some neighbourhood of the origin, we denote by $a_n = a_n(f)$ the coefficient of z^n in the Taylor expansion of f. By S we denote, as usual, the class of functions f holomorphic and univalent in U subjected to the normalization $a_0 = 0$, $a_1 = 1$. For any f holomorphic in U, its Dirichlet integral is denoted by

$$D(f) = \int_0^{2\pi} \int_0^1 |f'(re^{i\theta})|^2 r \, dr \, d\theta = \int_U |f'|^2 \, d\sigma ,$$

where $d\sigma$ denotes planar measure.

Let now α denote a fixed complex number, and let

$$S_\alpha = \{f \in S : a_2(f) = \alpha\}.$$

We are concerned with the problem of minimizing D(f) in the class S_α, in other words, minimizing the area of f(U) with respect to all $f \in S_\alpha$.

Let us make some preliminary observations. First of all, the class S_α is empty if $|\alpha| > 2$; secondly, if $|\alpha| = 2$, S_α consists only of the single function $\bar{\omega}k(\omega z)$ where $\omega = \alpha/2$ and $k(z) = z(1-z)^{-2}$ is the Koebe function, and this function has an infinite Dirichlet integral. Therefore, we only have a sensible problem when $|\alpha| < 2$. Also, since the transformation of passing from f(z) to $\bar{\lambda}f(\lambda z)$, where λ is a complex number of modulus one, leaves invariant both the class S and the value of the Dirichlet integral, there is no loss of generality in assuming $0 \le \alpha < 2$. Finally, observe that the problem is trivial for $0 \le \alpha \le \frac{1}{2}$, since for $f \in S_\alpha$ we have

$$D(f) = \pi \sum_{n=1}^{\infty} n|a_n|^2 \ge (1 + 2\alpha^2)\pi$$

and equality is attained here for the quadratic polynomial $f(z) = z + \alpha z^2$, which is readily seen to be univalent in U for $0 \le \alpha \le \frac{1}{2}$, and thus the unique solution of our minimum problem.

The existence of an extremal f is easily established, and we deduce a number of conditions which any extremal necessarily satisfies, as well as some properties of the quantity

$$A(\alpha) = \inf D(f), \quad f \in S_\alpha$$

in its dependence on α, in the interval $[0, 2)$. Every extremal domain V (that is, the image of the unit disc under the map given by an extremal f) is shown to be bounded, have a boundary of finite length, and satisfy remarkable quadrature identities. Moreover, in the presence of an assumption concerning the topological nature of an extremal domain which we are thus far unable to prove we are able to give the complete solution of the extremal problem.

Now we present our main results. First an argument based on 'interior variation' establishes

Theorem 1. Let $0 \le \alpha < 2$, and let $f \in S_\alpha$ be an extremal function, i. e. $D(f) = A(\alpha)$. Writing

$$f(z) = \sum_{n=1}^{\infty} a_n z^n \quad (a_1 = 1, \ a_2 = \alpha)$$

we have

$$\sum_{n=1}^{\infty} n^2 |a_n|^2 \le 2(2-a_2)^{-1} \sum_{n=1}^{\infty} n |a_n|^2.$$

Corollary. If f is an extremal function, f' is in the Hardy class H^2.

Corollary. Every extremal domain $V = f(U)$ is bounded, and its boundary consists of rectifiable arcs whose total length is finite.

Concerning the qualitative character of extremal domains, we can prove

Theorem 2. Let $\frac{1}{2} < \alpha < 2$. Then, every extremal domain V admits a non-empty set of boundary points, each of which fails to belong to the closure of any component of the complement of the closure of V.

Moreover, extremal domains satisfy a remarkable type of quadrature formula:

Theorem 3. *Let f be an extremal function in S_α where $0 \le \alpha < 2$ (that is, $D(f) = A(\alpha)$) and let $V = f(U)$ be the associated extremal domain. Then*

$$\int_V (h(w) - h(0) - h'(0)w)\,d\sigma = 0$$

for all functions h holomorphic in V such that

$$\left| \int_{w_1}^{w_2} h(w)\,dw \right| \le K |w_1 - w_2| \qquad (w_1, w_2 \in V)$$

where $K = K_h$ is a constant depending only on h. (The last integral is, of course, independent of the path of integration in G.)

Corollary. *If V is an extremal domain, then*

$$\int_V w^n\,d\sigma = 0, \quad n = 2, 3, \ldots .$$

A complement to this Corollary is

Theorem 4. *If V is an extremal domain, and $0 < \alpha < 2$, then*

$$0 < \operatorname{Re} \int_V w\,d\sigma \le (2(2 - \alpha))^{-1} \int_V d\sigma.$$

The complete solution of the problem will now be given, if certain topological conditions are assumed. Let $\tilde{V}$ denote the interior of the closure of V (i.e. $\tilde{V}$ is V 'with slits erased') and assume

(i) $\tilde{V}$ is simply connected;

(ii) every boundary point of $\tilde{V}$ lies in the closure of some component of the complement of the closure of $\tilde{V}$.

(*Note*: In a preliminary report of our work which has been circulated, we overlooked the fact that our demonstrations tacitly assumed condition (ii); we are grateful to Lars-Inge Hedberg for calling our attention to this lapse.)

Theorem 5. *If $\frac{1}{2} < \alpha < 2$, there is at most one extremal domain*

V for which (i) and (ii) above hold. Assuming one (and hence, exactly one) such V exists, then

(a) $\tilde{V}$ is a cardioid, the conformal image of U under a quadratic polynomial;

(b) V is $\tilde{V}$ minus a rectilinear slit (the cardioid and the position of the slit can be determined explicitly);

(c) the area of V is

$$A(\alpha) = (27/8)\,\pi\,(2 - \alpha)^{-2}.$$

An important tool in the proof of Theorem 5 is the following result characterizing simply connected domains over which holomorphic functions can be integrated by means of quadrature formulas of a certain simple type, a result which seems to be of independent interest:

Theorem 6. Let G be any simply connected domain of finite area in the complex w-plane. Suppose there exists a point $w_0 \in G$, and complex numbers λ_0, λ_1 such that for every g holomorphic and integrable over G

$$\int_G g(w)\,d\sigma = \lambda_0 g(w_0) + \lambda_1 g'(w_0).$$

Then, there exists a quadratic polynomial P(z), univalent in U and satisfying $P(0) = w_0$, such that G is the image of U under the map $w = P(z)$.

Conversely, if P(z) is any quadratic polynomial univalent in U, the image G of U under the map $w = P(z)$ satisfies the above identity for every g holomorphic and integrable over G, where $w_0 = P(0)$ and

$$(2.27) \qquad \lambda_0 = \int_G d\sigma, \quad \lambda_1 = \int_G (w - w_0)\,d\sigma.$$

This theorem can be greatly generalized, in the direction of a general theory of domains which admit finite quadrature formulas. However, our method requires essentially that G is simply connected, and this is at present the obstacle to our solution of the extremal problem. Were we able to prove Theorem 6 without topological hypotheses, we could make a corresponding improvement in Theorem 5, i. e. we should obtain the complete and explicit solution of the extremal problem.

Technion,
Haifa, Israel

Royal Institute of Technology,
Stockholm, Sweden

A REMARK ON SCHLICHT FUNCTIONS WITH QUASICONFORMAL EXTENSIONS

LARS V. AHLFORS*

1. Recently Schiffer and Schober have studied extremal problems for schlicht functions with a k-q.c. extension. Their results include a sharpening of the Grunsky inequalities. It is perhaps not without interest to observe that this particular result can be attained very quickly without use of variational techniques, simply by applying a basic property of quasiconformal reflection. Needless to say, the variational methods are of course essential for the deeper results of Schiffer and Schober.

We shall prove the following coefficient theorem:

Theorem. Let F be schlicht in $\Delta^* = \{|z| > 1\}$ with $F(\infty) = \infty$, and assume that F has a k-q.c. extension to $\Delta = \{|z| < 1\}$. Let Q_m be a polynomial of degree m, and define the coefficients c_n, $n \geqq -m$, by the expansion

$$(1.1)\quad Q_m[F(z)] = \sum_{n=-m}^{\infty} c_n z^{-n}.$$

Then

$$(1.2)\quad \sum_1^{\infty} n|c_n|^2 \leqq k \sum_1^{m} n|c_{-n}|^2 - (1-k)\left|\sum_1^{\infty} nc_n c_{-n}\right|,$$

and this implies

$$(1.3)\quad \left|\sum_1^{\infty} nc_n c_{-n}\right| \leqq k \sum_1^{m} n|c_{-n}|^2.$$

---

* Research partially supported under NSF Grant GP-38886.

2. We single out the special case $Q_m(w) = w$. With the usual normalization

$$F(z) = z + \sum_1^\infty b_n z^{-n}$$

we find:

Corollary 1. The area theorem is strengthened to

$$\sum_1^\infty n|b_n|^2 \leqq k - (1 - k)|b_1|$$

which implies $|b_1| \leqq k$. Also, the area of the complement of $F(\Delta^*)$ is at least equal to $(1 - k)(1 + |b_1|)\pi \geqq (1 - k)\pi$.

The Faber polynomials P_h associated with F are characterized by developments of the form

$$P_h[F(z)] = z^h + h \sum_{n=1}^\infty b_{hn} z^{-n}.$$

If the theorem is applied with

$$Q_m = \sum_{h=1}^m h^{-1} t_h P_h$$

one obtains

Corollary 2. The Golusin and Grunsky inequalities can be strengthened to

$$(2.1) \quad \sum_{n=1}^\infty n\left|\sum_h b_{hn} t_h\right|^2 \leqq k \sum_1^\infty n^{-1}|t_n|^2$$

and

$$(2.2) \quad \left|\sum_{h,n} b_{hn} t_h t_n\right| \leqq k \sum_1^\infty n^{-1}|t_n|^2.$$

For the sake of simplicity we have omitted a negative term on the right hand side of (2.1). Because of this simplification (2.2) is not an arithmetic consequence of (2.1); in fact, the Cauchy-Schwarz inequality would yield (2. 2) with $\sqrt{k}$ in the place of k. It is (2. 2) that was proved by Schiffer and Schober.

3. One of the earliest results on quasiconformal reflections is the following:

Lemma. _Let Ω and Ω^* be complementary Jordan regions in the extended complex plane, and assume that there exists a sense reversing quasiconformal mapping of Ω on Ω^* which leaves the boundary points fixed, and whose dilatation is everywhere $\leq K$. Let u and u* be harmonic functions in Ω and Ω^* with equal boundary values (if $\infty \in \Omega^*$ it is understood that $u^*(\infty)$ is finite). Then the Dirichlet integrals of u and u* satisfy $D(u^*) \leq KD(u)$._

The proof is a triviality. The hypothesis of the lemma is satisfied for $\Omega^* = F(\Delta^*)$ and $K = (1 + k)/(1 - k)$.

4. We apply the lemma to $\Omega^* = F(\Delta^*)$ and choose $u = \operatorname{Re} Q_m(w)$ in the complementary region Ω. The boundary values of u are given by

$$u \circ F = \operatorname{Re}\left[\sum_{-m}^{\infty} c_n z^{-n}\right]$$

for $|z| = 1$ (or by the radial limits). Clearly, u* will have the same boundary values if we define it by

$$u^* \circ F = \operatorname{Re}\left[\sum_{n=1}^{m} \bar{c}_{-n} z^{-n} + \sum_{n=0}^{\infty} c_n z^{-n}\right].$$

The Dirichlet integrals can be evaluated as line integrals, and it is well known that one obtains

$$D(u) = \sum_1^m n|c_{-n}|^2 - \sum_1^{\infty} n|c_n|^2$$

$$D(u^*) = \sum_1^m n|\bar{c}_{-n} + c_n|^2 + \sum_{m+1}^{\infty} n|c_n|^2.$$

It is found that the inequality $D(u^*) \leq KD(u)$ can be written in the form

$$\sum_1^\infty n|c_n|^2 \leqq k\sum_1^m n|c_{-n}|^2 - (1-k)\,\mathrm{Re}[\sum_1^m nc_n c_{-n}].$$

If Q_m is replaced by $e^{i\alpha}Q_m$ the expression $\sum nc_n c_{-n}$ acquires a factor $e^{2i\alpha}$. Therefore the last term can be replaced by the negative of the absolute value, and we have proved (1. 2).

5. In order to show that (1. 2) implies (1. 3) we introduce the notations

$$A = |\sum_1^m nc_n c_{-n}|$$

$$B = \sum_1^m n|c_{-n}|^2$$

$$C = \sum_1^\infty n|c_n|^2 .$$

Then $A^2 \leqq BC$ by the Schwarz-Cauchy inequality, and (1. 2) becomes $C \leqq kB - (1-k)A$. Hence $A^2 \leqq kB^2 - (1-k)AB$ or $(A-kB)(A+B) \leqq 0$ so that $A \leqq kB$ as asserted.

Harvard University,
Cambridge, U. S. A.

Added in proof: O. Lehto has informed me that he proved $\sum_1^\infty n|b_n|^2 \leq k$ in 1971, and that the same result had been obtained independently by R. Kühnau.

SOME EXTREMAL PROBLEMS FOR UNIVALENT FUNCTIONS, HARMONIC MEASURES, AND SUBHARMONIC FUNCTIONS

ALBERT BAERNSTEIN II

Let $u(z)$ be a subharmonic function in the annulus $r_1 < |z| < r_2$ $(0 \le r_1 < r_2 \le \infty)$. We define a new function u^* by

$$(1) \qquad u^*(re^{i\theta}) = \sup_E \int_E u(re^{i\psi})d\psi \qquad (0 \le \theta \le \pi)$$

where the sup is taken over all measurable sets $E \subset [0, 2\pi)$ with Lebesgue measure 2θ. This function was introduced by the author in [1]. From results established there one can obtain

Theorem A. u^* <u>is subharmonic in the semi-annulus</u>

$$\{z : r_1 < |z| < r_2, \ 0 < \arg z < \pi\}.$$

In this talk I will present some new results from several areas of complex function theory which have theorem A, or variations on it, as the main ingredient in their proof.

1. Integral means of univalent functions. Let S be the class of all functions f analytic and univalent in the unit disk with $f(0) = 0$, $f'(0) = 1$, and let

$$g(te^{i\phi}) = \frac{te^{i\phi}}{(1 - te^{i\phi})^2}$$

be the Koebe function.

Theorem 1. <u>Let</u> $A(r)$ <u>be a convex non-decreasing function of</u> $\log r$. Then for $f \in S$ and $0 < t < 1$,

$$\int_0^{2\pi} A(|f(te^{i\phi})|)d\phi \le \int_0^{2\pi} A(|g(te^{i\phi})|)d\phi .$$

In particular, we have for $0 < t < 1$,

$$\int_0^{2\pi} |f(te^{i\phi})|^p \, d\phi \leq \int_0^{2\pi} |g(te^{i\phi})|^p \, d\phi \qquad (0 < p < \infty), \tag{2}$$

$$\int_0^{2\pi} \log^+ |f(te^{i\phi})| \, d\phi \leq \int_0^{2\pi} \log^+ |g(te^{i\phi})| \, d\phi .$$

The best previously known result in the direction of (2) is apparently that of Bazilevic [4] who showed that (2) holds for $p = 1, 2$ if a universal constant is added to the right hand side.

Consider now any univalent function $f(te^{i\phi})$ in the unit disk. Let D be the conformal image of f in the z-plane, and let D^* be the circularly symmetrized domain of D. Let g be the conformal map of the unit disk onto D^* with $g(0) = |f(0)|$.

Theorem 2. With f and g as just described, the conclusion of theorem 1 holds.

Here is the idea of the proofs of theorems 1 and 2. Details will appear in [3]. Let F and G be the inverse functions of f and g and define $u = -\log|F|$, $v = -\log|G|$. Extend u and v to the whole z-plane by setting them equal to zero outside the images of f and g respectively. Then u is subharmonic in the plane, except for a logarithmic singularity at $f(0)$, and a similar statement holds for v.

It is not hard to show that theorems 1 and 2 are consequences of the inequality

$$\int_0^{2\pi} [u(re^{i\theta}) - \alpha]^+ d\theta \leq \int_0^{2\pi} [v(re^{i\theta}) - \alpha]^+ d\theta, \quad (0 < r < \infty, \ 0 < \alpha < \infty),$$

and it is easily seen that this inequality is implied by the inequality

$$u^*(re^{i\theta}) \leq v^*(re^{i\theta}) \qquad (0 < r < \infty, \ 0 < \theta < \pi), \tag{3}$$

where u^* and v^* are defined by (1). The key to proving (3) with the situation of theorem 1 is the fact that the proposed extremal function v^* is harmonic in the upper half plane, whereas, by theorem A, u^* is subharmonic there. Similarly, with the situation of theorem 2, it turns out that

$$v^* + 2\pi \log^+ \frac{r}{|f(0)|}$$

(or v^*, in case $f(0) = 0$) is harmonic in the part of D^* that lies in the upper half plane.

2. Integral means of harmonic measures. Consider now a subdomain D of the unit disk $|z| < 1$ whose boundary has a non-empty intersection γ with $|z| = 1$. Let D^* again denote the circular symmetrization of D. Let

$$u(z) = \omega(z, \gamma, D), \quad v(z) = \omega(z, \gamma^*, D^*),$$

and set $u = 0$, $v = 0$ at points inside the unit disk but outside D and D^* respectively.

Theorem 3. Let $B(x)$ be a convex non-decreasing function of x. Then for $0 < r < 1$,

$$\int_0^{2\pi} B(u(re^{i\theta}))d\theta \le \int_0^{2\pi} B(v(re^{i\theta}))d\theta .$$

By considering $B(x) = x^p$ and letting $p \to \infty$ we obtain as a corollary

$$\sup_\theta u(re^{i\theta}) \le v(r). \tag{4}$$

Inequality (4) provides a solution to the Carleman-Milloux problem which is more refined than the one commonly known (see, e. g. [11, p. 94]). For domains D that are simply connected and contain the origin (4) can be deduced from a theorem of Haliste [8], but our results are valid for any D.

To prove theorem 3 one first shows that the desired integral inequality is a consequence of (3) (for $0 < r < 1$), where now u and v are harmonic measures instead of Green's functions. The proof of (3) proceeds along the same lines as in the proof of theorem 2.

3. Extremal problems for subharmonic functions. The results discussed below were originally stated in terms of entire or meromorphic functions, f, but they are still valid if one replaces $\log |f|$ by (depending on the context) any subharmonic function or difference of subharmonic functions.

Let u be an unbounded subharmonic function in the plane. Set

$$M(r) = \sup_\theta u(re^{i\theta}) .$$

The author has recently proved the following theorem [2].

Theorem 4. Let β and λ be positive numbers satisfying $\beta \leq \pi$, $\beta\lambda < \pi$. Then either (a) there exist arbitrarily large values of r for which

$$u(re^{i\theta}) > M(r) \cos \beta\lambda$$

holds on a θ-interval of length at least 2β, or else

(b) $$\lim_{r\to\infty} r^{-\lambda} \log M(r)$$

exists, and is positive or infinite.

For $\beta = \pi$ this is Kjellberg's improvement [10] of the $\cos \pi\rho$ theorem. Using $\beta = \frac{\pi}{2}$, $\lambda = 1$, it is easy to deduce the regularity theorem of Heins [9] about functions subharmonic in a half plane with bounded boundary values. The proof of theorem 4 is based on considerations involving two variants of the function u*.

The original application of (an extended version of) theorem A was to provide a proof of the 'spread relation' [1], which enabled Edrei [5] to give a solution to the 'deficiency problem' for meromorphic functions with lower orders between $\frac{1}{2}$ and 1. We mention also that theorem A, together with Edrei's concept of 'Pólya peaks', can be used to give short proofs of Paley's conjecture (proved first by Govorov [7]) and the Edrei-Fuchs 'ellipse theorem' [6, theorem 1].

One of the most interesting unsolved extremal problems of the above sort is the 'k(λ) problem': for λ not an integer, determine

$$\sup\left(\liminf_{r\to\infty} \frac{\int_0^{2\pi} u^+(re^{i\theta})d\theta}{\int_0^{2\pi} u(re^{i\theta})d\theta}\right)$$

where the sup is taken over all subharmonic functions u of order λ. For $0 < \lambda < 1$ a special case of the ellipse theorem asserts that

$$u_\lambda(re^{i\theta}) = r^\lambda \cos \lambda\theta \ (|\theta| \leq \pi)$$

is the extremal function, and it is conjectured that $(-1)^{[\lambda]}u_\lambda(re^{i\theta})$ is extremal in the general case. The author thinks that perhaps an effective way to approach this problem would be to look for more elaborate auxiliary functions of the u^* type.

References

1. A. Baernstein, Proof of Edrei's spread conjecture, Proc. London Math. Soc. (3) 26 (1973), 418-34.
2. A. Baernstein, A generalization of the $\cos \pi\rho$ theorem, Trans. Amer. Math. Soc., 191 (1974).
3. A. Baernstein, Inequalities for integral means of univalent functions and harmonic measures, to appear.
4. I. E. Bazilevic, On distortion theorems and coefficients of univalent functions (Russian), Mat. Sb. N. S., 28, 70 (1951), 147-64.
5. A. Edrei, Solution of the deficiency problem for functions of small lower order, Proc. London Math. Soc., (3) 26 (1973), 435-45.
6. A. Edrei and W. H. J. Fuchs, The deficiencies of meromorphic functions of order less than one, Duke Math. J., 27 (1960), 233-50.
7. N. V. Govorov, On Paley's problem, (Russian), Funk. Anal. 3 (1969), 35-40.
8. K. Haliste, Estimates of harmonic measures, Ark. Mat., 6 (1965), 1-31.
9. M. Heins, On the Phragmén-Lindelöf principle, Trans. Amer. Math. Soc., 60 (1946), 238-44.
10. B. Kjellberg, A theorem on the minimum modulus of entire functions, Math. Scand., 12 (1963), 5-11.
11. R. Nevanlinna, Eindeutige Analytische Funktionen, Berlin, 1936.

Washington University,
St. Louis, U. S. A.

ON COEFFICIENT PROBLEMS FOR CERTAIN POWER SERIES

D. A. BRANNAN

1. Introduction

In [2] the problem was raised of verifying the family of inequalities

$$(1.1)\quad \left(\frac{1 + xz}{1 - z}\right)^{\alpha} \ll \left(\frac{1 + z}{1 - z}\right)^{\alpha},$$

where $\alpha \geq 1$, $|x| = 1$; this was settled in [1]. Here we provide a shorter proof of (1.1), together with various similar inequality families, and discuss to what extent the results are best-possible. Throughout we will use the definitions and notation of [2].

2. The main result

Our fundamental result is

Theorem 1. Suppose that $|x| = 1$, $\alpha > 0$, $\varepsilon > 0$, $\alpha + \frac{1}{2}\varepsilon \geq 1$. Then

$$(2.1)\quad \frac{(1 + xz)^{\alpha}}{(1 - z)^{\alpha+\varepsilon}} \ll \frac{(1 + z)^{\alpha}}{(1 - z)^{\alpha+\varepsilon}},$$

with equality for each nth coefficient $(n \geq 1)$ if and only if $x = 1$.

Proof. (i) Let

$$f_x(z) \equiv \frac{(1 + xz)^{\alpha}}{(1 - z)^{\alpha+\varepsilon}} = \sum_{n=0}^{\infty} A_n(x)z^n, \qquad A_0(x) = 1.$$

Then, for $m \geq 1$,

$$A_m(x) = \sum_{r=0}^{m} \frac{(\alpha+\varepsilon)(\alpha+\varepsilon+1)\ldots(\alpha+\varepsilon+m-r-1)}{(m-r)!}\binom{\alpha}{r}x^r,$$

so that the triangle inequality shows that

$$(2.2)\quad f_x(z) \underset{n}{\ll} f_1(z) \qquad (\alpha \geq n).$$

Also let

$$F_x(z) = \left(\frac{1 + xz}{1 - z}\right)^{\alpha}, \qquad F_x(0) = 1;$$

as above we see that

$$(2.3) \quad F_x(z) \underset{n}{\ll} F_1(z) \qquad (\alpha \geq n).$$

(ii) For $\beta \geq 1$ and $0 < \alpha < \frac{\pi}{2}$, let $P(\alpha, \beta)$ denote the class of functions $P(z)$ where

$$P(z) = [p(z)]^{\beta},$$

where $p(z)$ is analytic in $|z| < 1$, $p(0) = P(0) = 1$, and

$$\alpha - \frac{\pi}{2} < \arg p(z) < \alpha + \frac{\pi}{2} \qquad (|z| < 1).$$

Then, if $x = e^{2i\alpha}$, Theorem 2.2 of [2] shows that $P(z)$ has the integral representation

$$P(z) = \int_0^{2\pi} \left(\frac{1 + xze^{it}}{1 - ze^{it}}\right)^{\beta} d\mu(t),$$

for some increasing function $\mu(t)$ on $[0, 2\pi]$ with $\mu(2\pi) - \mu(0) = 1$. Consequently the set of extreme points of $P(\alpha, \beta)$ can be shown to be

$$\text{Ext } P(\alpha, \beta) = \left\{\left(\frac{1 + xze^{it}}{1 - ze^{it}}\right)^{\beta}, \qquad 0 \leq t < 2\pi\right\}.$$

Furthermore, for $|x| = 1$, $\beta \geq 1$, and any δ with $0 < \delta < \beta$, we have, by standard extreme point arguments, that the nth coefficient of $(1 + xz/1 - z)^{\beta-\delta}(1 + z/1 - z)^{\delta}$ is majorised by that of $(1 + y_n z/1 - z)^{\beta}$, for some y_n with $|y_n| = 1$.

(iii) Fix an integer $n \geq 1$. Suppose that $|x| = 1$, $\alpha > 0$, $\varepsilon > 0$. Then we have that

$$(2.5) \quad \frac{(1 + xz)^{\alpha}}{(1 - z)^{\alpha+\varepsilon}} = \left[\left(\frac{1 + xz}{1 - z}\right)^{\alpha}\left(\frac{1 + z}{1 - z}\right)^{\frac{1}{2}\varepsilon}\right]\frac{1}{(1 - z^2)^{\frac{1}{2}\varepsilon}}.$$

By (ii) the nth coefficient of the function in the square bracket is majorised by $(1 + y_n z/1 - z)^{\alpha+\frac{1}{2}\varepsilon}$ since $\alpha + \frac{1}{2}\varepsilon \geq 1$, for some y_n with $|y_n| = 1$; and by (2.3) this is majorised up to the nth coefficient by $(1 + z/1 - z)^{\alpha+\frac{1}{2}\varepsilon}$

as long as $\alpha + \frac{1}{2}\varepsilon \geq n$. Hence from (2.5) we deduce that (2.2) holds with the weaker restriction that $\alpha + \frac{1}{2}\varepsilon \geq n$.

(iv) We now turn to the restriction in (2.3). By applying the result of (iii) in the form

$$\frac{(1 + xz)^{\alpha-1}}{(1 - z)^{\alpha+1}} \underset{n-1}{\ll} \frac{(1 + z)^{\alpha-1}}{(1 - z)^{\alpha+1}} \qquad (\alpha \geq n-1)$$

(i.e. with $\varepsilon = 2$), we can deduce that

$$\frac{\partial}{\partial z}[F_x(z)] \equiv \alpha(1 + x)\frac{(1 + xz)^{\alpha-1}}{(1 - z)^{\alpha+1}}$$

$$\underset{n-1}{\ll} \frac{\partial}{\partial z}[F_1(z)],$$

and so (by integration) that

$$F_x(z) \underset{n}{\ll} F_1(z) \qquad (\alpha \geq n-1);$$

in other words (2.3) holds with the weaker restriction that $\alpha \geq n-1$.

(v) We now repeat the procedures of (iii) and (iv). At successive stages we obtain (2.2) and (2.3) with the lower bound for α reduced by 1. After $(n - 1)$ steps we thus obtain that $|A_n(x)| \leq A_n(1)$ for $\alpha + \frac{1}{2}\varepsilon \geq 1$. Since this is valid for each particular n, (2.1) follows. The equality assertion is easily verified.

A special case of Theorem 1 that is of some interest is

Theorem 2. Suppose that $|x| = 1$, $\alpha \geq \frac{1}{2}$. Then

$$\frac{(1 + xz)^{\alpha}}{(1 - z)^{\alpha+1}} \ll \frac{(1 + z)^{\alpha}}{(1 - z)^{\alpha+1}}$$

with equality for each nth coefficient $(n \geq 1)$ if and only if $x = 1$.

We now establish the main result of Aharonov and Friedland [1], namely

Theorem 3. Suppose that $|x| = 1$, $\alpha \geq 1$. Then

$$\frac{(1 + xz)^{\alpha}}{1 - z} \ll \frac{(1 + z)^{\alpha}}{1 - z},$$

with equality for each nth coefficient $(n \geq 1)$ if and only if $x = 1$.

Proof. We note first that it is sufficient to prove the result for $1 < \alpha < 2$. Let

$$f_x(z) = \frac{(1 + xz)^{\alpha}}{1 - z} \qquad (|x| = 1).$$

Then

$$\frac{\partial}{\partial z}[f_x(z)] = \frac{(1 + xz)^{\alpha-1}}{(1 - z)^{3-\alpha}} \cdot \frac{(1 + \alpha x) + x(1 - \alpha)z}{(1 - z)^{\alpha-1}}$$

$$= g_x(z) \,.\, h_x(z) \quad \text{(say)}.$$

By Theorem 1, $g_x(z) \ll g_1(z)$. Further

$$\frac{1}{\alpha-1} \cdot \frac{\partial}{\partial z}[h_x(z)] = \frac{1}{(1-z)^{\alpha-1}} \cdot \frac{(1+\alpha x) - \alpha xz + xz^2}{1 - z}$$

$$= \frac{1}{(1-z)^{\alpha-1}}[(1+\alpha x) + z + (1+x)\frac{z^2}{1-z}]$$

$$\ll \frac{1}{\alpha-1} \cdot \frac{\partial}{\partial z}[h_1(z)] ;$$

consequently $h_x(z) \ll h_1(z)$, and so $f_x(z) \ll f_1(z)$. The equality assertion is easily verified.

Our next result, (1.1), has already been proved as a stepping-stone in the proof of Theorem 1:

Theorem 4. <u>Suppose that</u> $|x| = 1$, $\alpha \geq 1$. <u>Then</u>

$$\left(\frac{1 + xz}{1 - z}\right)^{\alpha} \ll \left(\frac{1 + z}{1 - z}\right)^{\alpha},$$

<u>with equality for each</u> n^{th} <u>coefficient</u> $(n \geq 1)$ <u>if and only if</u> $x = 1$.

3. Some applications

First of all we remark that Theorem 4 solves the coefficient conjectures for (a) the class of functions of bounded boundary rotation in $|z| < 1$ [8] and (b) the class of close-to-convex functions of order β $(\beta > 0)$ in $|z| < 1$ [6]; for details, see [2, pp. 10-11].

Next we observe how Theorem 1 can be used to settle the case of equality in the arc-length problem for close-to-convex functions in $|z| < 1$ [3][4].

Theorem 5. Let $f(z) = z + a_2 z^2 + \dots$ be close-to-convex in $|z| < 1$, and $k(z) = z/(1-z)^2$. Then, for $0 < r < 1$ and $\lambda \geq 1$,

$$\int_0^{2\pi} |f'(re^{i\theta})|^\lambda d\theta \leq \int_0^{2\pi} |k'(re^{i\theta})|^\lambda d\theta ,$$

with equality if and only if $f(z)$ is a rotation of $k(z)$.

Proof. We deal first with the case $\lambda = 1$. Following [3] we know that it is sufficient to show that

$$(3.1) \quad I(\alpha, t) \leq I(0, 0)$$

where

$$I(\alpha, t) = \int_0^{2\pi} \frac{|1 + re^{i\theta} e^{2i\alpha}|}{|1 - re^{i\theta}|\,|1 - r^{i\theta} e^{it}|^2} d\theta ,$$

$-\pi/2 < \alpha < \pi/2$, $0 \leq t < 2\pi$, with equality in (3.1) if and only if $\alpha = t = 0$.

We apply the identity

$$(3.2) \quad \int_0^{2\pi} |\sum_{n=1}^{\infty} nb_n r^n e^{in\theta}|^2 d\theta = 2\pi \sum_{n=1}^{\infty} n^2 |b_n|^2 r^{2n}$$

to the function

$$\sum_{n=0}^{\infty} b_n z^n = [\frac{(1 + ze^{i\alpha})(1 + ze^{it})}{(1 - z)(1 - ze^{it})}]^{\frac{1}{2}} \frac{1}{1 - z^2 e^{2it}}$$

$$= g(z) \,.\, h(z) \qquad \text{(say).}$$

By a (trivial) modification of the Herglotz formula, we see that $g(z) \ll \dfrac{1 + z}{1 - z}$, and so

$$(3.3) \quad \sum_{n=0}^{\infty} b_n z^n \ll \frac{(1 + z)^{\frac{1}{2}}}{(1 - z)^{3/2}} .$$

(3.2) and (3.3) together yield (3.1).

Using Theorem 1 in place of the Herglotz formula, the result can be proved in a similar way for $\lambda \geq 1$. The case of equality is easily checked.

Similarly we may prove the following result, which, for $k > 4$, is a generalisation of a result of Eenigenburg [5].

Theorem 6. Let $f(z) = z + a_2 z^2 + \ldots \in V_k$, and

$$f_k(z) = \frac{1}{k}\left[\left(\frac{1+z}{1-z}\right)^{k/2} - 1\right].$$

If $0 < r < 1$, $k \geq 4$, and $\lambda \geq 4/k$, then

$$\int_0^{2\pi} |f'(re^{i\theta})|^\lambda d\theta \leq \int_0^{2\pi} |f_k'(re^{i\theta})|^\lambda d\theta,$$

with equality if and only if $f(z)$ is a rotation of $f_k(z)$.

4. Discussion of the main results

It is natural to ask whether Theorems 1, 2, 3, and 4 are valid for $\alpha \geq 0$, $\varepsilon \geq 0$. We now show that in general they are not, and raise some interesting problems in this connection.

First of all, the standard major-minor arc method [7, p. 108] shows that, if

$$(4.1) \qquad \sum_{n=0}^{\infty} A_n(x) z^n = \frac{(1+xz)^\alpha}{(1-z)^\beta}, \qquad A_0(x) = 1,$$

where $\alpha > 0$, $\beta > 0$, and $|x| = 1$, then for a fixed x with $x \neq 1$ there exists an $n_0 = n_0[x, \alpha, \beta]$ such that

$$(4.2) \qquad |A_n(x)| < A_n(1)$$

for all $n > n_0$.

It is of some interest to see whether (4.2) holds for $\alpha + \beta < 2$ in (4.1); in view of Theorem 1, (4.2) holds for $\alpha + \beta \geq 2$. In general (4.2) turns out to be false, but it is true in certain special cases. Since

$$|A_1(x)| = |\alpha x + \beta| \leq A_1(1),$$

the first interesting coefficient is $A_2(x)$. To investigate this, we use

Lemma 7. Suppose that $a > 0$, $b > 0$, $|x| = 1$, and $p(x) = a + x - bx^2$. Then $|p(x)| \leq |p(1)|$ with equality only for $x = 1$ if and only if $4ab \leq a - b$.

Proof. With $c = \cos\theta$, we have

$$|p(e^{i\theta})|^2 = (a^2 + 2ab + b^2 + 1) + 2c(a - b) - 4abc^2,$$

from which the result follows.

With Lemma 7 as our main tool, we can, without much effort, verify

Theorem 8. *Let $A_n(x)$ be defined by (4.1), and $0 < \alpha < 1$. Then*

(i) *If $\beta = 1$, $\max_{|x|=1} |A_2(x)| = A_2(1)$ if and only if $\frac{1}{2}(\sqrt{17} - 3) \le \alpha < 1$;*

(ii) *if $\beta = \alpha + 1$, $\max_{|x|=1} |A_2(x)| = A_2(1)$ if and only if $\frac{1}{2}(\sqrt{3} - 1) \le \alpha < 1$; and*

(iii) *if $\alpha = \beta$, $\max_{|x|=1} |A_2(x)| \le A_2(1)$ if and only if $1/\sqrt{2} \le \alpha < 1$.*

In view of the complication involved in considering $A_n(x)$ in (4.1) with the two parameters α and β, we now prove several results only for the special case $\beta = \alpha + 1$ of (4.1).

Theorem 9. *Let*

$$\sum_{n=0}^{\infty} A_n(x)z^n = \frac{(1 + xz)^\alpha}{1 - z}, \qquad A_0(x) = 1,$$

where $0 < \alpha < 1$, $|x| = 1$. Then

(i) *$|A_3(x)| \le A_3(1)$ with equality if and only if $x = 1$;*

(ii) *there exists an n_α such that*

$$\max_{|x|=1} [\mathrm{Re}\, A_{2n}(x)] > A_{2n}(1)$$

for all $n > n_\alpha$;

(iii) *for a given α and a given $n > 1$, there exists a neighbourhood N of $\theta = 0$ where*

$$\max_{\theta \in N} [\mathrm{Re}\, A_{2n+1}(e^{i\theta})] = A_{2n+1}(1);$$

and

(iv) *for each given $n >$ some absolute constant n_0, there exists an*

α_n and a neighbourhood $N(n)$ of $\theta = 0$ such that

$$\max_{\theta \in N(n)} |A_n(e^{i\theta})| = A_n(1)$$

for all α with $\alpha_n < \alpha < 1$.

Proof. (i) The proof is similar to that of Theorems 7 and 8; the detailed calculation involves considering the cases $0 < \alpha < \frac{1}{2}$, $\frac{1}{2} \leq \alpha \leq \frac{2}{3}$, $\frac{2}{3} < \alpha < 1$ separately.

We conjecture that (i) holds for $A_3(x)$ in (4.1) for all $\alpha > 0$ and $\beta > 0$; a study of

$$A_3(x) = \binom{\beta}{3} + \binom{\beta}{2}\binom{\alpha}{1}x + \binom{\beta}{1}\binom{\alpha}{2}x^2 + \binom{\alpha}{3}x^3$$

using a computer suggests that this conjecture is likely to be true. We further conjecture that $|A_{2n+1}(x)| \leq A_{2n+1}(1)$ for $n \geq 1$ for all $\alpha > 0$, $\beta > 0$, or at least for all $\alpha > 0$ with $\beta = 1$; numerical calculation again makes this seem likely.

(ii) With $x = e^{i\theta}$, we have

$$\begin{aligned} \mathrm{Re}[A_{2n}(e^{i\theta})] &= \sum_{r=0}^{2n} \binom{\alpha}{r} \cos(r\theta) \\ &> A_{2n}(1) \end{aligned}$$

for some θ sufficiently near to 0 if

$$\alpha(-\tfrac{1}{2}) + \binom{\alpha}{2}(-\tfrac{1}{2} \cdot 2^2) + \ldots + \binom{\alpha}{2n}(-\tfrac{1}{2} \cdot [2n]^2) > 0,$$

(4.3) i.e. $\displaystyle\sum_{r=1}^{2n} \binom{\alpha}{r} r^2 < 0.$

We write this as $\displaystyle\sum_{r=1}^{n} a_r < 0$, where we put

$$\begin{aligned} a_r &= \binom{\alpha}{2r}(2r)^2 + \binom{\alpha}{2r-1}(2r-1)^2 \\ &= \binom{\alpha}{2r-1}[2r(\alpha-1) + 1]. \end{aligned}$$

Then (4.3) can be written as

(4.4) $2(1 - \alpha)S_n > T_n$,

where

$$S_n = \sum_{r=1}^{n} r\binom{\alpha}{2r-1}, \quad T_n = \sum_{r=1}^{n} \binom{\alpha}{2r-1}.$$

We now show that $T_n \to$ a finite limit as $n \to \infty$, and so is bounded for all n, and that $S_n \to \infty$ as $n \to \infty$. Our result will then follow from (4.4).

All the terms in T_n are positive, so that T_n is an increasing function of n. Since $\sum_{n=0}^{\infty} \binom{\alpha}{n} z^n = (1+z)^{\alpha}$ is absolutely convergent on $|z| = 1$ for $\alpha > 0$, we have that $\sum_{r=0}^{\infty} (-1)^r \binom{\alpha}{r} = 0$ and $\sum_{r=0}^{\infty} \binom{\alpha}{r} = 2^{\alpha}$, and so $0 < T_n < 2^{\alpha-1}$ (< 1).

By Gauss' Test [9, p. 463] we can show that the series of positive terms $\sum_{r=1}^{n} (2r-1)\binom{\alpha}{2r-1}$ is divergent, and so $S_n \to \infty$ as $n \to \infty$. This completes the proof of (ii).

In fact by estimating T_n more carefully we can show that $(1-\alpha)\log(n_{\alpha})$ remains bounded as $\alpha \to 1$.

(iii) We note first that, using a similar method to that in (ii), it is sufficient to show that

$$\frac{2(1-\alpha)\sum_{r=1}^{n} r\binom{\alpha}{2r-1}}{\binom{\alpha}{2n+1}(2n+1)^2} = \frac{A_n}{B_n} \to \tfrac{1}{2} \tag{4.5}$$

as $n \to \infty$. Using the major-minor arc method as in [7, p. 108] we have that

$$A_n \sim \frac{\alpha 2^{1-\alpha}}{\Gamma(1-\alpha)} n^{1-\alpha}, \quad B_n \sim \frac{\alpha 2^{-\alpha}}{\Gamma(1-\alpha)} n^{1-\alpha}$$

as $n \to \infty$; (4.5) then follows.

(iv) Using a method similar to that of (ii), but with $|A_n(e^{i\theta})|^2$ in place of $\operatorname{Re} A_n(e^{i\theta})$, we see that it is enough to prove that

$$\left[\sum_{r=0}^{n} r\binom{\alpha}{r}\right]^2 \le \sum_{r=0}^{n} \binom{\alpha}{r} \cdot \sum_{r=0}^{n} r^2\binom{\alpha}{r} \tag{4.6}$$

for all $\alpha < 1$ with $\alpha >$ some $\alpha_n > 0$. Since we are only interested in

α near 1, we write (4.6) in the form

$$(4.7) \quad O\{(1-\alpha)^2\} \le \alpha - (1-\alpha)S_2 - 2(1-\alpha)S_3 + 2(1-\alpha)S,$$

where

$$S_1 = \sum_{r=2}^{n} \frac{(-1)^r}{r-1}, \quad S_2 = \sum_{r=2}^{n} \frac{(-1)^r}{r(r-1)}, \quad S_3 = \sum_{r=2}^{n} \frac{r}{r-1}(-1)^r.$$

Since S_1, S_2, S_3 are bounded for all n, (4.7) follows at once.

Note. When $\beta = \alpha$ in (4.1), it is easy to verify also that

$$|A_3(x)| \le A_3(1)$$

for $|x| = 1$, $0 < \alpha < 1$, with equality only if $x = 1$.

The author wishes to thank S. Fairthorne for his assistance in the computer investigations, and S. Friedland for a stimulating conversation.

References

1. D. Aharonov, S. Friedland. On an inequality connected with the coefficient conjecture for functions of bounded boundary rotation, to appear.
2. D. A. Brannan, J. G. Clunie, W. E. Kirwan. On the coefficient problem for functions of bounded boundary rotation, Ann. Acad. Sci. Fenn. Ser. Al Math., (1973), no. 523, 18 pp.
3. J. G. Clunie, P. L. Duren. An arc-length problem for close-to-convex functions, Addendum, J. London Math. Soc., 41 (1966), 181-2.
4. P. L. Duren. An arc-length problem for close-to-convex functions, J. London Math. Soc., 39 (1964), 757-61.
5. P. J. Eenigenburg, to appear.
6. A. W. Goodman. On close-to-convex functions of higher order, to appear.
7. W. K. Hayman. Multivalent functions, Cambridge (1968).

8. V. Paatero. Über die konforme Abbildung von Gebieten deren Ränder von beschränkter Drehung sind, Ann. Acad. Sci. Fenn. Ser. A, 9 (1931), 77 pp.

9. R. A. Rankin. Mathematical Analysis, Academic Press (1963).

Queen Elizabeth College,
London, England.

APPROXIMATION BY ANALYTIC FUNCTIONS UNIFORMLY CONTINUOUS ON A SET

LEON BROWN and ALLEN SHIELDS

Abstract

Let D denote the open unit disc in the complex plane. Let $F \subset D$ be a relatively closed subset. Let $U(F)$ denote the class of functions that are holomorphic in D and uniformly continuous on F. Let F^- denote the closure of F. The functions in $U(F)$ may be extended by continuity to be defined on F^-. We use the norm

(1) $$\|f\|_F = \sup\{|f(z)| : z \in F\}.$$

If J is a compact set in the plane, then $A(J)$ denotes the class of functions continuous on J and holomorphic in the interior of J.

Problem. _Describe the class of functions, defined on_ F^-, _that can be approximated uniformly on_ F^- _by functions in_ $U(F)$. _Equivalently, describe the closure of_ $U(F)|_{F^-}$ _in_ $C(F^-)$.

To answer this question we require the notion of the hull of F with respect to the family $U(F)$.

$$\text{Hull}(F) = F^- \cup \{z \in D : |f(z)| \le \|f\|_F, \text{ for all } f \in U(F)\}.$$

It is easy to see that $\text{Hull}(F)$ is a compact set.

Theorem. _The closure of_ $U(F)$ _in_ $C(K)$ _is precisely_ $A(\text{Hull}(F))$.

The proof is based in part on Banach algebra methods. It works for more general domains D than the unit disc. For example, it is sufficient if each boundary point of D is a boundary point of a component of $\mathbf{C} \setminus D^-$.

Question 1. Is the theorem valid in all bounded plane domains D?

Question 2. Describe the closure of the class $B(F)$ in $C_b(F)$.

Here $B(F)$ denotes the class of those holomorphic functions in D that are bounded on F, and $C_b(F)$ denotes the bounded continuous functions on F (not necessarily uniformly continuous) with the norm (1).

These questions are somewhat related to theorems of Arakelian.

Wayne State University,
Detroit, U. S. A.

University of Michigan,
Ann Arbor, U. S. A.

A MEROMORPHIC FUNCTION WITH ASSIGNED NEVANLINNA DEFICIENCIES

DAVID DRASIN[1]

1. Statement of result

Theorem 1. *Let non-negative numbers* δ_i, θ_i $(i = 1, \ldots, N \leq \infty)$ *be given such that*

$$0 < \delta_i + \theta_i \leq 1 \qquad (i = 1, \ldots, N),$$

$$\sum_{i=1}^{N} (\delta_i + \theta_i) \leq 2,$$

together with a sequence $\{a_i\}$ $(i = 1, \ldots, N)$ *of distinct extended complex numbers. Then there exists a function* f *which is meromorphic in the finite plane with*

$$\delta(a_i) = \delta_i, \quad \theta(a_i) = \theta \qquad (i = 1, \ldots, N)$$

and

$$\delta(a) = \theta(a) = 0 \qquad (a \notin \{a_i\}).$$

Finally, let $\phi(r)$ *be a positive increasing function with*

$$\phi(r) \to \infty \qquad (r \to \infty). \tag{1.1}$$

Then f *may be chosen so that, in addition, its Nevanlinna characteristic satisfies*

$$T(r) < r^{\phi(r)} \tag{1.2}$$

1 Research partially supported by National Science Foundation (U. S. A.).

for sufficiently large r.

Here we use the standard notations of Nevanlinna theory; for example, $\theta(a)$ is the index of multiplicity (Verzweigungsindex) of a. The function f thus provides a solution to the 'inverse problem' of the Nevanlinna theory (cf. [2], Ch. 7, §5).

The problem has been the subject of several earlier studies, of which we note Nevanlinna [4], Goldberg (cf. [2], p. 547 and p. 550), and Fuchs-Hayman (cf. [3], §4.1).

The general solution to the inverse problem must be of infinite order (cf. [6]); (1.2) shows that this may be as 'small' an infinite order as desired.

It is a pleasure to acknowledge many useful conversations with Allen Weitsman, and the influence of our earlier study [1].

Here, we shall only outline the methods used to obtain a function with assigned deficiencies, although the theory of §2 applies to the full inverse problem. Full proofs will appear elsewhere.

The letter A will be used to denote an absolute constant, usually different at different occurrences. If A depends on a parameter p, it will be written A(p).

Our approach is to reduce the problem to constructing a suitable quasi-conformal mapping, and $g(z)$ or, in §3, $\gamma(w)$ will refer to such a quasi-conformal map. We reserve the notation $f(\zeta)$ for a meromorphic function defined in the finite ζ-plane.

2. On a principle of Teichmüller

Let g be a sense-preserving homeomorphism from the finite complex plane **C** to a (necessarily simply-connected open) Riemann surface $\mathfrak{F}$ which covers the sphere; thus we may view, ambiguously, $g(z)$, for each $z \in \mathbf{C}$, as an extended complex number or a point on $\mathfrak{F}$. Suppose there exist open sets $D_1, D_2, \ldots$ with $D_i \cap D_j = \emptyset$ $(i \neq j)$, $\cup \overline{D_i} = \mathbf{C}$ such that every bounded subset of **C** meets at most finitely many D_i. On each D_i we assume that g has continuous first partial derivatives with respect to x and y, and one-sided partial derivatives on ∂D_i. Then if $z \in D_i$ and $g(z) \neq \infty$, let $g_{\bar{z}} = \frac{1}{2}(g_x + ig_y)$, $g_z = \frac{1}{2}(g_x - ig_y)$ (if $g(z) = \infty$, consider g^{-1} instead) and let

$$(2.1) \quad d_g(z) = |g_{\bar{z}}/g_z| ;$$

$d_g(z)$ is the <u>dilatation</u> of g at z. If $z \in \partial D_i$ for some i we define (2.1) by taking $d_g(z)$ as the supremum of all possible expressions $|g_{\bar{z}}/g_z|$, where the one-sided derivatives at z are considered.

Unless $d_g \equiv 0$, g is not a meromorphic function, but since the range of g is a Riemann surface the expressions

$$n(r, a, g), \quad \bar{n}(r, a, g) \qquad (a \in \mathbf{C} \cup \{\infty\})$$

are readily defined. We shall relate these to the value-distribution functionals of some meromorphic function f by imposing conditions on the dilatation (2.1).

Suppose first that g is quasi-conformal, i.e.

$$(2.2) \quad d_g(z) < k < 1 \qquad (z \in \mathbf{C}),$$

where k is some fixed constant never to change in this paper. Then according to Teichmüller [5] the surface $\mathfrak{F}$ is parabolic, and so there is a meromorphic function $f(\zeta)$ mapping the finite ζ-plane conformally onto $\mathfrak{F}$. We define

$$(2.3) \quad \psi(\zeta) = g^{-1}(f(\zeta)),$$

which maps the ζ-plane homeomorphically onto the z-plane. It is convenient to normalise f so that (2.4) and (2.7) below will hold.

First, we shall require that

$$(2.4) \quad \psi(0) = 0;$$

then for $\rho > 0$ the circle $|\zeta| = \rho$ corresponds under ψ to a curve Γ_ρ in the z-plane which encircles $z = 0$. Let

$$(2.5) \quad r_1(\rho) = \inf\{\, |z| : z \in \Gamma_\rho \},\quad r_2(\rho) = \sup\{\, |z| : z \in \Gamma_\rho \},$$

$$(2.6) \quad \omega(\rho) = \log\{\, r_2(\rho)/r_1(\rho) \},$$

and, finally, set

$$(2.7) \quad r_1(1) = 1.$$

Theorem 2. Let g be a quasi-conformal mapping whose dilatation satisfies (2.2), and let ψ be given by (2.3), with (2.4) and (2.7). Then

$$(2.8) \quad r_2(\rho) < \rho^{A_1}, \quad r_1(\rho) > \rho^{A_2} \qquad (\rho > 1).$$

Further, to each $\varepsilon > 0$ and $\rho_0 > 1$ corresponds a $\sigma > 0$ such that

$$(2.9) \quad \omega(\rho) < \varepsilon \qquad (\rho > \rho_0)$$

if

$$(2.10) \quad D_g(r) \equiv \int_0^{2\pi} d_g(re^{i\theta})d\theta < \sigma \qquad (r > \rho_0^A).$$

Teichmüller and Goldberg required

$$(2.11) \quad \iint_{|z|>1} d_g(z)|z|^{-2}dxdy < \infty$$

rather than

$$(2.12) \quad D_g(r) \to 0 \qquad (r \to \infty),$$

which is anticipated by (2.10). In [1] Weitsman and I proposed (2.12) as a more convenient condition for the Nevanlinna theory. The form of our construction, given below, suggests that (2.11) will rarely occur for functions considered here.

It is clear from (2.6) that

$$(2.13) \quad n(r_1(\rho), a, g) \leq n(\rho, a, f) \leq n(r_2(\rho), a, g) \qquad (a \in \mathbf{C} \cup \{\infty\}),$$

where the extreme terms of (2.13) reflect the quasi-conformal function g.

Choose a b^* which is to satisfy

$$(2.14) \quad n(r, b^*, g) \sim \exp\{\int_1^r \lambda(t)t^{-1}dt\} \quad (r \to \infty),$$

where $\lambda(t)$ increases very slowly to infinity. In order that the composition (2.3) transfer the data of g to f, (2.13) yields that necessarily

$$\text{(2.15)} \quad \int_{r_1(\rho)}^{r_2(\rho)} \lambda(t)t^{-1}dt \to 0 \qquad (\rho \to \infty).$$

The construction of §3 will produce numbers $R_1 < R_1' < R_2 < R_2' < \ldots$ with R_n'/R_n and R_{n+1}/R_n' tending rapidly to infinity. When $R_n < t < R_n'$ we will have $\lambda(t) = n$, and λ increases from n to $n+1$ in the interval $R_n' < t < R_{n+1}$. According to (2.8) there is a number S_n with the property that if $r_2(\rho) > R_n'$, then necessarily $\rho > S_n$, and so (2.8), (2.9) and (2.10) yield that there are constants $\sigma_n > 0$, A with the property that if

$$\text{(2.16)} \quad D_g(r) \le \sigma_n \qquad (r > (R_n')^A),$$

then

$$\text{(2.17)} \quad \omega(\rho) < (n+2)^{-3} \qquad (\rho > S_n).$$

Thus if $R_n' \le r_2(\rho) \le R_{n+1}$, then

$$\lambda(r_2(\rho))\omega(\rho) \le (n+1)(n+2)^{-3} < n^2,$$

and so (2.15) holds. (It is not hard to see that (2.15) is also a sufficient condition that the Nevanlinna data of g transfer to f; the specific choice of b^* is not essential.)

This discussion shows that it is enough to construct a quasi-conformal mapping such that the mean dilatation (2.10) decreases very rapidly with respect to the growth of $\lambda(t)$ in (2.14). By increasing the ratios R_n'/R_n, R_{n+1}/R_n' the requirement (2.16) is more readily met, and the growth of $g(z)$, as measured by $\lambda(|z|)$, is retarded. The relation (2.8) now implies that by suitably augmenting these ratios (1.2) may be fulfilled.

3. A meromorphic function with assigned deficiencies

Given the sequences $\{\delta_i\}$, $\{a_i\}$ with $\infty \notin \{a_i\}$, define a function $\Delta(a)$ by

$$\Delta(a_i) = \delta_i \qquad (i = 1, 2, \ldots)$$
$$\Delta(a) = 0 \qquad (a \notin \{a_i\}).$$

Next, consider a sequence $\ldots, b_{-2}, b_{-1}, b_0, b_1, \ldots$ of complex numbers with $b_{-j} = b_j$, $b_j \neq b_{j+1}$ such that if

$$|E_n(a)| = \text{card}\{ j : b_j = a, \ -(n-1) \le j \le n \},$$

then

$$(3.1) \quad n^{-1}|E_n(a)| \to \Delta(a) \qquad (n \to \infty)$$

for all a. The number b^*, introduced in (2.14), is taken disjoint from the $\{b_j\}$.

For each integer j, the methods of [2] (p. 492) yield a map γ_j from the half plane $P = \{ \mathcal{R}e\, w \ge 0 \}$ $(w = u + iv = se^{it})$ to a bordered Riemann surface such that for all $\delta > 0$

$$(3.2) \quad \gamma_j(w) \sim b_j + e^{-w} \qquad (w \to \infty, \ -\tfrac{1}{2}\pi + \delta \le t \le \tfrac{1}{2}\pi - \delta),$$

and on the line $\mathcal{R}w = 0$,

$$(3.3) \quad \gamma_j(iv) = \gamma_{j+1}(-iv) \qquad (v > 0).$$

Further,

$$(3.4) \quad d_{\gamma_j}(w) \le 2 \qquad (\mathcal{R}e\, w \ge 0),$$

$$(3.5) \quad \int_{-\pi/2}^{\pi/2} d_{\gamma_j}(w)dt \to 0 \qquad (s = |w| \to \infty, \ t = \arg w).$$

Our function g is defined in stages; after stage n it is defined 'appropriately' in $\{ |z| \le R_n \}$. That is, if n is a positive integer we assume that there exist $\{R_k\}_{k=1}^{n}$, $\{R'_k\}_{k=1}^{n-1}$, $\{\sigma_k\}_{k=1}^{n-1}$ so that $R_1 < R'_1 < \ldots < R_{n-1} < R'_{n-1} < R_n$, g is defined in $\{ |z| \le R_n \}$ and (3.6)-(3.10) below hold. First, with ϕ given in (1.1), A_1, A_2 in (2.8), and λ in (2.14) we require that

$$(3.6) \quad \lambda(r^{A_1}) < \phi(r^{A_2}) \qquad (r \le R_n).$$

Also, for all extended complex a we have

(3.7) $\frac{n(r, a, g)}{n(r, b^*, g)} = \Delta(a) + \varepsilon(r, a)$

with

(3.8) $|\varepsilon(r, a)| < \frac{2}{k} \qquad (R_{k-1} \le r \le R_k,\ k = 2, \dots, n).$

The dilatation of g is required to satisfy

(3.9) $D_g(r) \le \sigma_k \qquad (r > (R_k')^A,\ k = 1, \dots, n-1),$

where σ_k is chosen in accord with (2.16), (2.17). Finally, we need rather precise information about $g(z)$ when $|z|$ is nearly R_k $(k = 1, \dots, n)$. If

$$A_k = \{ \tfrac{1}{2}R_k \le |z| \le R_k \} \qquad (k = 1, \dots, n),$$

then A_k is divided into $2k$ sectors $J_{-(k-1)}, \dots, J_k$ each of angular opening π/k. Let $e^{i\alpha_j}z^k$ map J_j into P for an appropriate real α_j. Then there is a positive constant C, which depends only on n, so that g may be defined by

(3.10) $g(z) = \gamma_j(Ce^{i\alpha_j}z^k) \qquad (z \in A_k \cap J_j;\ k = 1, \dots, n,\ -k \le j \le k).$

(In particular, $g(z)$ is very close to b_j on almost all of $A_k \cap J_j$.)

It is enough to choose σ_n, R_n', R_{n+1} so that (3.6)-(3.10) hold with $(n + 1)$ in place of n. The discussion of §2, notably (2.16) and (2.17), yields a suitable σ_n, and then R_n'/R_n, R_{n+1}/R_n' are chosen large, where the magnitudes of the ratios will depend on σ_n, the numbers b_n and b_{n+1}, and the need to conserve (3.6).

First, let g be extended for $|z| \ge R_n$ by (3.10) (where now $k = n$). Then (3.5) yields an R_n^* $(> R_n)$ with

(3.11) $D_g(r) < \sigma_n \qquad (r > R_n^*),$

and we have thus augmented the domain of g so that

(3.12) $g(z) = \gamma_j(Ce^{i\alpha_j}z^n) \quad (z \in J_j,\ -(n-1) \le j \le n,\ R_n \le |z| \le R_n^*).$

Next we choose a suitable $R_n' > R_n^*$. To be comparable with (2.16), we want $R'_n > (R^*_n)^A$, but the choice also depends on σ_n, b_n, b_{n+1}. When $-(n-1) \le j \le n-1$, g is simply defined by (3.12) with R'_n in place of R^*_n; a similar although more complicated formula is used in $J_n \cap \{ R^*_n \le |z| \le R'_n \}$ so that (3.11) holds and our advance to stage $(n+1)$ will be more convenient.

It remains to choose R_{n+1} and define g for $R'_n \le |z| \le R_{n+1}$. First let $-(n-1) \le j \le (n-1)$. We shall arrange that $J_j \cap \{ |z| = r \}$ $(R'_n < r < R_{n+1})$ be an arc whose angular opening decreases slowly from π/n to $\pi/(n+1)$. Indeed we can define $J_j \cap \{R'_n < |z| < R_n\}$ as the preimage of a map

$$(3.13) \quad h_j : J_j \cap \{ R'_n < |z| < R_{n+1} \} \to P \cap \{ W_n < |w| < W'_n \}$$

for suitable W_n, W'_n. The formulas are simplest after some changes of variables: $x = \log r = \log |z|$, $x_n = \log R'_n$, $x_{n+1} = \log R_{n+1}$ and, in (3.13), $h_j(re^{i\theta}) = H_j(x + i\theta)$ with $H_j = \exp(S + iT)$,

$$(3.14) \quad \begin{aligned} S &= n\left(x + \int_{x_n}^{x} \{ m(t) + p(t) \} dt\right) + C \\ T &= n\theta(1 + m(x)) + \beta_j(x) + \alpha_j \end{aligned}$$

with C an appropriate constant. In (3.14), m and β_j are linear functions of x such that

$$m(x_n) = \beta_j(x_n) = 0, \quad m(x_{n+1}) = -(n+1)^{-1}, \quad \beta_j(x_{n+1}) = \beta_j(\text{const.}),$$
$$p(x_n) = 0, \quad |p(t)| < (x_{n+1} - x_n)^{-1}.$$

Simple computations yield

$$(3.15) \quad \begin{aligned} S_x &= n(1 + m(x) + p(x)), \quad T_\theta = n(1 + m(x)), \\ S_\theta &= 0, \quad T_x = n\theta m'(x) + \beta'_j(x). \end{aligned}$$

Thus if $x_{n+1} - x_n$ is sufficiently large (i.e. if R_{n+1}/R_n is sufficiently large) our mapping h_j will have as small a dilatation as desired in $J_j \cap \{ R'_n < |z| < R_{n+1} \}$, and consequently if R_{n+1}/R_n is large we define g in

$$(3.16)\quad \{R'_n < |z| < R_{n+1}\} \cap \{\bigcup_{-(n-1)}^{n-1} \overline{J_j}\}$$

by, if $z \in \overline{J_j}$,

$$(3.17)\quad g(z) = \gamma_j(h_j(z)),$$

where γ_j is described in (3. 2)-(3. 5) and h_j in (3. 13), (3. 14).

Finally we define g in the complement of the set (3. 16) with respect to $\{R'_n < |z| < R_{n+1}\}$. This is a sector $\mathcal{T}_n$ such that $\mathcal{T}_n \cap \{|z| = r\}$ is an arc whose angular measure increases from π/n to $3\pi/(n+1)$ as r increases from R'_n to R_{n+1}. We have $\mathcal{T}_n \cap \{|z| = R'_n\} = J_n$ and shall view $\mathcal{T}_n \cap \{|z| = R_{n+1}\}$ as being 'split' into arcs of J_{n+1}, J_n and J_{-n}.

To do this we proceed in a manner analogous to that discussed at the beginning of this section. Thus we will map $\mathcal{T}_n \cap \{R'_n < |z| < R_{n+1}\}$ into a region P^* of the w $(= se^{it})$ plane by $w = h^*(z)$; P^* is described in (3. 22) below. If R_{n+1}/R_n is sufficiently large, this may be achieved with dilatation as small as desired. In P^* is defined a quasi-conformal map γ^* of small dilatation and, finally, g is extended to $\mathcal{T}_n \cap \{R'_n < |z| < R_{n+1}\}$ by

$$(3.18)\quad g(z) = \gamma^*(h^*(z)).$$

Near the boundary of P^*, the ideas of [2] let us define γ^* so that it has small dilatation and the definitions (3. 17) and (3. 18) agree on $\partial\mathcal{T}_n$. However, (3. 2) must be replaced by a more subtle behaviour for γ^*. Let L be a Möbius transformation with $L(\infty) = b_n$, $L(0) = b_{n+1}$. Then away from ∂P^*, γ^* is defined as a composition

$$(3.19)\quad \gamma^*(w) = L\psi F(w),$$

where ψ has preassigned small dilatation and F is the entire function given in the

Proposition. <u>Given</u> $\sigma' > 0$, $\eta > 0$ <u>there exist</u> r_0, σ_1 <u>and</u> M <u>such that if</u> $r_1 > r_0$, $r_2 > Mr_1$ <u>there is an entire function</u> $F(w)$ $(w = se^{it})$ <u>with zeros on the positive axis and</u>

(3.20) $\log F(w) = -(1 + h(w))e^{-i\pi\Lambda(s)}w^{\Lambda(s)}C(s),$

where $s|C'(s)|/|C(s)| < \sigma', \quad C(s) > 0,$

(3.21) $|h(w)| < \sigma' \qquad (s \geq r_0, \quad \eta \leq |t| \leq \pi)$

and $\Lambda(s)$ increases with $\Lambda(s) = 1 \ (s \leq 2r_1)$, $\Lambda(s) = 2 \ (s \geq \frac{1}{2}r_2)$.

(If we ask that Λ increase from $1 + \varepsilon$ to $2 - \varepsilon$ this is very simple; (cf. [2], Ch. 2, §5).)

For an appropriate choice of r_1, r_2 we let

(3.22) $P^* = \{w = se^{it} : r_1 \leq s \leq r_2, \quad |t| < \pi - \pi/(2\Lambda(s))\},$

and introducing ψ in (3.19) allows us to suppress the small term $h(w)$ in (3.20); the dilatation of ψ may be made as small as desired by choosing σ' in (3.21) sufficiently small.

When $\Lambda(s) < \frac{3}{2}$, F is large on nearly all of $\{|w| = s\} \cap P^*$, but as s increases and $\Lambda(s)$ passes $\frac{3}{2}$ a component where F is very near to 0 develops on $P^* \cap \{|w| = s\}$ about the positive w-axis. As s increases so that $\Lambda(s)$ approaches 2, this component becomes an arc of opening $\frac{1}{2}\pi$ in $P^* \cap \{|w| = s\}$.

The simple form of (3.20) and the definitions (3.17), (3.18) and (3.19) ensure that it is easy to satisfy (3.6)-(3.10), and thus (3.1) ensures that our quasi-conformal function g will have the preassigned deficiencies Hence, according to the discussion of §2, so will some meromorphic function $f(\zeta)$.

References

1. D. Drasin and A. Weitsman. Meromorphic functions with large sums of deficiencies, to appear in Advances in Mathematics.
2. A. Goldberg and I. Ostrowski. The distribution of values of meromorphic functions (in Russian), Nauka, Moscow (1970).
3. W. K. Hayman. Meromorphic functions, Oxford (1964).
4. R. Nevanlinna. Über Riemannsche Flächen mit endlich vielen Windungspunkten, Acta Math. 58 (1932), 295-373.

5. O. Teichmüller. Untersuchungen über konforme und quasikonforme, Abbildung, Deutsche Math., 3 (1938), 621-78.

6. A. Weitsman. A theorem on Nevanlinna deficiencies, Acta Math., 128 (1972), 41-52.

Imperial College,
London, England.

Purdue University,
Lafayette, U. S. A.

ESTIMATION OF COEFFICIENTS OF UNIVALENT FUNCTIONS BY A TAUBERIAN REMAINDER THEOREM

PETER L. DUREN

Let S be the class of functions

$$f(z) = z + a_2 z^2 + a_3 z^3 + \dots$$

analytic and univalent in the unit disk. Some time ago, W. K. Hayman showed that for each $f \in S$, the sequence $\{|a_n|/n\}$ converges to a limit $\alpha \leq 1$, with equality occurring only for a rotation of the Koebe function. The first step in Hayman's proof is the observation that each $f \in S$ has a direction $e^{i\theta}$ of maximal growth, in the sense that

$$\lim_{r \to 1} (1 - r)^2 |f(re^{i\theta})| = \alpha ,$$

and the limit is 0 for every other direction. The second and more difficult step is then the deduction that $|a_n|/n \to \alpha$.

I. M. Milin has recently simplified this second step in the case $\alpha > 0$. His argument is based upon the following result, which may be viewed as a new Tauberian theorem. Let

$$g(r) = \sum_{n=0}^{\infty} b_n r^n, \qquad b_0 = 1,$$

be a power series with complex coefficients, convergent for $-1 < r < 1$. Let

$$s_n = \sum_{k=0}^{n-1} b_k, \qquad \sigma_n = \frac{1}{n} \sum_{k=0}^{n-1} s_k ,$$

and let

$$\log g(r) = \sum_{n=1}^{\infty} c_n r^n .$$

Milin's Theorem. Suppose that $|g(r)| \to \alpha$ as $r \to 1$, and $\sum_{n=1}^{\infty} n|c_n|^2 < \infty$. Then $|s_n| \to \alpha$ and $|\sigma_n| \to \alpha$.

Hayman's theorem is deduced by taking $\theta = 0$ and by considering

$$g(r) = \frac{(1-r)^2}{r} f(r), \quad \text{for which } \sigma_n = \frac{a_n}{n} .$$

The Tauberian condition $\Sigma n|c_n|^2 < \infty$ follows (for $\alpha > 0$) from a recent theorem of Bazilevich.

In a similar way, by appealing to a Tauberian remainder theorem essentially due to Freud and Korevaar, one can establish the following quantitative version of Hayman's theorem.

Theorem. Suppose that $f \in S$ satisfies the inequality

$$(*) \qquad \left| \frac{(1-r)^2}{r} f(r) - \lambda \right| \leq B(1-r)^{\delta}, \quad 0 < r < 1,$$

for some complex number $\lambda \neq 0$ and for some positive constants B and δ. Then

$$\left|\frac{a_n}{n} - \lambda\right| \leq \frac{C}{\log n} ,$$

where C depends only on $|\lambda|$, B, and δ.

Corollary. For each $f \in S$ satisfying (*) for some λ, $0 < \alpha = |\lambda| < 1$, the coefficients satisfy $|a_n| < n$ for all $n \geq N$, where N depends only on α, B and δ.

University of Michigan,
Ann Arbor, U. S. A.

THE PADÉ TABLE OF FUNCTIONS HAVING A FINITE NUMBER OF ESSENTIAL SINGULARITIES

ALBERT EDREI

Let

$$a_0 + a_1 z + a_2 z^2 + \dots = f(z) \qquad (a_0 \neq 0)$$

have a positive radius of convergence.

With every integer $k \geqq 0$, we associate an ordered pair $(m(k), n(k))$ of non-negative integers and denote by $R_{mn}(z)$ the rational function which occupies the nth row and m^{th} column of the Padé table. By definition

$$R_{mn} = \frac{P_m}{Q_n} \qquad (Q_n(z) \not\equiv 0),$$

where the polynomial P_m (of degree $\leqq m$) and the polynomial Q_n (of degree $\leqq n$) satisfy the extremal conditions which characterize the table. The existence and uniqueness of the functions R_{mn} are well-known and will be taken for granted [2; pp. 235-7].

We study the behaviour of the Padé table for $f(z)$ <u>quasi-meromorphic and of finite order</u> Λ. This means that

I. $f(z)$ <u>is single-valued; its only singularities are poles and essential singularities</u>

$$\alpha_0, \alpha_1, \dots, \alpha_s \qquad (s < +\infty),$$

<u>which may be limit-points of poles.</u> [<u>One of the</u> α's <u>may be</u> ∞.]

II. <u>The order of</u> α_j <u>is</u> λ_j <u>and</u>

$$\lambda_0 + \lambda_1 + \lambda_2 + \dots + \lambda_s = \Lambda < +\infty.$$

Theorem. <u>Let</u> $f(z)$ <u>be quasi-meromorphic and of finite order</u> Λ. <u>Assume that</u> $m(k) > 0$ <u>and</u> $n(k) > 0$ <u>are such that</u>

(i) $\chi k^{\beta} < m + n \quad (k = 1, 2, 3, \ldots, \chi > 0)$,

for some fixed χ and

$$\beta > \frac{\Lambda}{2} ;$$

(ii) $m = o(n \log n), \quad n = o(m \log m)$,

as $k \to +\infty$.

Then, given $\rho > 0$ and $\delta > 0$, it is possible to find a measurable set Γ such that

$$\text{meas } \Gamma < \delta ,$$

and such that

$$R_{mn}(z) \to f(z) \quad (k \to +\infty) ,$$

uniformly for all z restricted by the conditions

$$|z| \leqq \rho , \quad z \notin \Gamma .$$

The case

$$\Lambda < 2 , \quad m = n = k,$$

is of particular interest because the Theorem then implies pointwise convergence of the whole diagonal of the Padé table, almost everywhere in the complex plane.

The introduction of the exceptional set Γ cannot be avoided unless the class of functions under consideration is severely restricted. This may be seen as follows.

Consider the Hankel determinants

$$A_n = \begin{vmatrix} a_n & a_{n-1} & a_{n-2} & \cdots & a_1 \\ a_{n+1} & a_n & a_{n-1} & \cdots & a_2 \\ a_{n+2} & a_{n+1} & & \cdots & a_3 \\ \cdots & \cdots & & \cdots & \cdots \\ a_{2n-1} & \cdots & & \cdots & a_n \end{vmatrix} \quad (n = 1, 2, 3, \ldots).$$

For the class of quasi-meromorphic functions of finite order Λ, the author has established [1; pp. 11-30] the 'best possible' inequality

$$\limsup_{n\to\infty} |A_n|^{1/n^2 \log n} \leqq e^{-1/\Lambda}, \tag{1}$$

and shown that little can be said about A_{n+1}/A_n beyond the obvious consequence of (1):

$$\liminf_{n\to\infty} |A_{n+1}/A_n|^{1/2n\log n} \leqq e^{-1/\Lambda}.$$

On the other hand, the arguments which lead to the proof of the Theorem, show that, if $\Lambda < 2$ and $A_n \neq 0$ $(n \geqq 1)$, the uniformity of the convergence of $\{R_{nn}(z)\}_{n=1}^{\infty}$, in some neighbourhood of the origin, requires

$$\limsup_{n\to\infty} |A_{n+1}/A_n|^{1/n\log n} \leqq e^{-1/\Lambda}.$$

The above condition has little significance from the function-theoretical point of view. Within the class of quasi-meromorphic functions of order Λ, it represents as serious a restriction as would be the requirement that, within the class of entire functions of finite order (for which $\limsup_{n\to\infty} |a_n|^{1/n\log n} < 1$), we consider only those for which $\limsup_{n\to\infty} |a_{n+1}/a_n|^{1/\log n} < 1$.

References

1. A. Edrei. Sur les déterminants récurrents et les singularités d'une fonction donnée par son développement de Taylor, Compositio Math., 7 (1939), 20-88.
2. O. Perron. Die Lehre von den Kettenbrüchen, 3rd ed., vol 2, B. G. Teubner, Stuttgart, 1957.

Syracuse University,
Syracuse, U. S. A.

EXTREMAL PROBLEMS OF THE COS $\pi\rho$-TYPE

ALBERT EDREI

Let $\mathfrak{F}_\mu$ be the class of meromorphic functions of lower order exactly equal to $\mu < +\infty$ and $\mathcal{E}_\mu$ the subclass formed by the entire members of $\mathfrak{F}_\mu$.

Consider, with their usual meaning, the well known notations of Nevanlinna's theory

$$(1) \qquad \log M(r, f),\ n(r, f),\ N(r, f),\ T(r, f),$$

to which we add

$$(2) \qquad \log m^*(r, f) = \min_{|z|=r} \log|f(z)|,$$

and the deficiency $\delta(\tau, f)$ of the value τ.

Wiman's relation

$$(3) \qquad \limsup_{r\to\infty} \frac{\log m^*(r, f)}{\log M(r, f)} \geq \cos \pi\mu \qquad (f \in \mathcal{E}_\mu,\ \mu < 1),$$

with μ replaced by the order ρ, has been classical for half a century. It is natural to consider, more generally, relations such as

$$(4) \qquad \limsup_{r\to\infty} \frac{G(r, f)}{H(r, f)} \geq \Lambda_{G/H}(\mu) \qquad (f \in \mathfrak{F}_\mu),$$

where G and H denote simple combinations of suitable functions in (1) and (2).

With this meaning of the symbols, (4) poses a problem which may be said to be of the $\cos \pi\rho$-type. The restrictions on μ (or to $f \in \mathcal{E}_\mu$) are sometimes due to the nature of the question; they are more frequently dictated by the complexity of the situation. For instance, a suitable extension of

$$(5) \qquad \limsup_{r\to\infty} \frac{N(r, f) + N(r, 1/f)}{T(r, f)} \geq \sin \pi\mu \qquad (f \in \mathfrak{F}_\mu,\ \tfrac{1}{2} \leq \mu < 1),$$

has never been obtained for $\mu > 1$.

We consider now some specific choice of $G(r, f)$, $H(r, f)$ and examine the associated problem of the cos $\pi\rho$-type.

I. The 'direct problem' is to be considered as 'solved' when the best possible value of $\Lambda_{G/H}(\mu)$ has been determined for all μ in a suitable range.

II. A function $f \in \mathfrak{F}_\mu$ is 'extremal' for the problem under consideration if equality holds in the corresponding relation (4).

Determination of the extremal functions

What can be said about the extremal functions thus defined (even their existence is not obvious)?

Is there a general method for attacking such problems?

The author proposes to describe an approach that leads to a satisfactory answer in all the following cases.

A. Wiman's relation (3);

B. Valiron's relation

$$\limsup_{r\to\infty} \frac{N(r, 1/f)}{\log M(r, f)} \geq \frac{\sin \pi\mu}{\pi\mu} \quad (f \in \mathcal{E}_\mu,\ \mu < 1);$$

C. Paley's relation

$$\limsup_{r\to\infty} \frac{T(r, f)}{\log M(r, f)} \geq \frac{1}{\pi\mu} \qquad (f \in \mathcal{E}_\mu,\ \tfrac{1}{2} \leq \mu);$$

D. The relation (5) and its generalization involving the ellipse

$$u^2 + v^2 - 2uv \cos \pi\mu = \sin^2 \pi\mu \quad (u = 1 - \delta(0, f),\ v = 1 - \delta(\infty, f)).$$

There are a few other cases in which the method is successful, in particular the treatment of

$$\limsup_{r\to\infty} \frac{\log m^*(r, f)}{T(r, f)} \geq \chi \frac{\pi\mu}{\sin \pi\mu} \qquad (f \in \mathfrak{F}_\mu,\ \mu < \tfrac{1}{2}),$$

where

$$\chi = \delta(\infty, f) - 1 + \cos \pi\mu > 0 .$$

Some common features of all the extremal functions which have been determined are worth noticing:

(i) there always exists a sequence $\{r_m\}$ of Pólya peaks of order μ, such that, for every $\sigma > 0$,

$$\lim_{m\to\infty} \frac{T(\sigma r_m)}{T(r_m)} = \sigma^{\mu} ; \tag{6}$$

(ii) outside exceptional disks, with sum of radii $o(r_m)$, the asymptotic behaviour of $\log|f(z)|$ is determined in annuli

$$\sigma^{-1} r_m \leq |z| \leq \sigma r_m \quad (m = 1, 2, 3, \ldots); \tag{7}$$

the parameter $\sigma > 1$ may be chosen arbitrarily large;

(iii) in the annuli (7), the asymptotic behaviour of the arguments of two values (which we take to be 0 and ∞) conforms to some simple pattern characteristic of the problem. ['Almost radial' distribution of zeros and poles is frequent.] The presence of $o(T(r_m))$ exceptional zeros and poles does not affect the asymptotic evaluations and cannot be precluded;

(iv) the local character of the preceding assertions cannot be replaced by a global one. In particular, (6) need not hold in the more complete form $T(\sigma r) \sim \sigma^{\mu} T(r)$ $(r \to \infty)$.

Syracuse University,
Syracuse, U. S. A.

A COS $\pi\lambda$-PROBLEM AND A DIFFERENTIAL INEQUALITY

MATTS ESSÉN

Let u be a non-constant subharmonic function in the complex plane $\underline{C}$. Let $\psi : [0, \infty) \to [0, 1)$ be a lower semicontinuous function such that

$$(1) \qquad m(r) \leq \cos \pi\psi(r) M(r), \quad r > 0,$$

where $M(r) = \max u(z)$, $|z| = r$, and $m(r) = \inf u(z)$, $|z| = r$.

What can we say about the growth of $M(r)$ as $r \to \infty$? It has been conjectured by B. Kjellberg that if (1) holds

$$(2) \qquad M(r) \leq \text{Const. } M(R) \exp\{-\int_r^R \psi(t)dt/t\}, \quad 0 < r < R.$$

This conjecture is known to be true in the following special cases. Let λ be given, $0 < \lambda < 1$.

(i) $\psi(r) = \lambda$, $r > 0$. Then (2) is a weak form of the Hellsten-Kjellberg-Norstad inequality (cf. [6]; this paper will in the sequel be referred to as HKN).

(ii) Let E be a union of intervals in $[0, R]$ and assume that

$$(3) \qquad m(r) \leq \cos \pi\lambda M(r), \quad r \in E.$$

The case $\lambda = \frac{1}{2}$ is due to A. Beurling (1933). The case $0 < \lambda < 1$ is in J. Lewis [7].

Unfortunately, it does not always follow from (1) that (2) is true. A counterexample can be found in Essén ([3], §6).

We can prove the following result. In this context, it is no essential restriction to assume that u is harmonic in the unit disc so that $u(0)$ is finite. We can also assume that $u(0) \leq 0$.

We assume that there exists a positive number μ such that the range of ψ is contained in the set $\{0\} \cup [\mu, 1)$.

Theorem 1. Let u and ψ be as above. Let the smallest positive number in the range of ψ be μ. Let

$$\Psi(r) = \mu\{(1 - \cos \pi\psi(r))/(1 - \cos \pi\mu)\}^{\frac{1}{2}}.$$

If (1) is true, there exists an absolute constant and a number r_0 (which only depends on u) such that

$$M(r) \leq \text{Const.} \{|u(0)| + M(R)\exp\{-\int_r^R \Psi(t)dt/t\}, \quad r_0 < r < R. \tag{4}$$

As a first application, consider the theorem of J. Lewis quoted above. We define $\psi(r) = \{\begin{matrix}\lambda, & r \in E, \\ 0, & r \notin E.\end{matrix}$ The smallest positive number in the range of ψ is λ, we see that $\Psi = \psi$ and it follows from (4) that

$$M(r) \leq \text{Const.} \{|u(0)| + M(R)\exp\{-\lambda m_l E(r, R)\}\}, \quad r_0 < r < R, \tag{5}$$

which is the result of Lewis. Here $m_l E(r, R) = \int_{E\cap(r, R)} dt/t$.

As a second application, we can obtain certain results of P. D. Barry ([2], Theorems 1 and 2).

The author and J. Lewis have extended Theorem 1 to the situation considered by A. Baernstein [1] (cf. Essén [3], §8). Let, if α and β are given positive numbers

$$m_\beta(r, \alpha, u) = \inf_{|\phi|<\beta} u(re^{i(\phi+\alpha)}),$$

$$\sup_\alpha m_\beta(r, \alpha, u) = m_\beta(r, u) = m_\beta(r).$$

Theorem 2. Let u be as above, let $R > 0$ be given, and let $\psi : [0, R] \to [0, \infty)$ be a lower semicontinuous function which is bounded on $[0, R]$. Let β be given such that

$$0 < \beta \leq \pi, \qquad 0 \leq \beta\psi(r) < \pi, \qquad 0 \leq r \leq R.$$

Let the smallest positive value in the range of ψ be μ. We define $\Psi_\beta(r) = \mu\{(1 - \cos \beta\psi(r))/(1 - \cos \beta\mu)\}^{\frac{1}{2}}$. If

$$m_\beta(r) \leq \cos \beta\psi(r)M(r), \qquad 0 \leq r \leq R, \tag{1'}$$

then (with constants as in Theorem 1)

(4') $$M(r) \le \text{Const.} \left(|u(0)| + M(R) \exp\{ - \int_r^R \Psi_\beta(t) dt/t \} \right), \quad r_0 < r < R.$$

Corollary 1. Let E be a union of intervals in $[0, R]$. Let β and λ be given such that $0 < \lambda < \infty$, $0 < \beta \le \pi$, $\beta\lambda < \pi$. If

$$m_\beta(r) \le \cos \beta\lambda M(r), \quad r \in E,$$

then (5) holds.

This is an extension of the theorem of Lewis quoted above.

It is easy to state and prove extensions of the result of P. D. Barry mentioned above. As an example, let us consider the following corollary which answers in part a question of A. Baernstein [1]: can certain results of Drasin and Shea be extended to this more general situation? Let ρ_0 be the order of u and let $\overline{\Lambda}E = \limsup_{R\to\infty} m_l E(1, R)/\log R$ be the upper logarithmic density of the set E.

Corollary 2. If β is given, $0 < \beta \le \pi$, $0 < \beta\rho_0 < \pi$, and

$$\limsup_{r\to\infty} m_\beta(r)/M(r) \le \cos \beta\rho_0,$$

there exists a set G, $\overline{\Lambda}G = 0$, such that

$$m_\beta(r)/M(r) \to \cos \beta\rho_0, \quad r \to \infty, \quad r \notin G.$$

Remark. The original result of P. D. Barry is in Theorem 1 in [2].

The main tool in the proof of Theorem 1 is the following theorem. By a solution of the differential equations or inequalities discussed here, we mean a function Φ such that Φ and Φ' are absolutely continuous, Φ'' is the a. e. existing derivative of Φ' and the equation or the inequality is satisfied a. e.

Theorem 3. Let $p : (-\infty, 0] \to [0, \infty)$ be a lower semicontinuous function which is uniformly bounded on each compact interval. Assume that there exists a solution of the inequality

(6) $\Phi''(t) - p(t)^2\Phi(t) \geq 0, \quad -\infty < t \leq 0,$

such that $\Phi(0) = 1, \ \lim_{t \to -\infty} \Phi(t) \leq 0.$

Let t_0 *be given,* $t_0 < 0$. *If* $\inf_t p(t) > 0$, *there exists a non-negative solution* Φ^* *of the equation*

(7) $\Phi^{*''} - p^*(t)^2\Phi^*(t) = 0 ,$

such that $\Phi^*(0) = 1, \ \Phi^*(-\infty) = 0,$ *and*

(8) $\Phi(t_0) \leq \Phi^*(t_0).$

Here p^* *is a measure preserving, non-decreasing re-arrangement of* p *on* $[t_0, 0]$ *and* $p^*(t) = \inf_{s<0} p(s)$ *on* $(-\infty, t_0)$.

If $\inf_s p(s) = 0$, *the statement above will still be true except that the conclusion* $\Phi^*(-\infty) = 0$ *is replaced by* $\Phi^*(t) = \Phi^*(t_0), \ t < t_0$.

Remark 1. It is fairly easy to obtain an estimate of $\Phi^*(t_0)$. Inequality (8) gives us than an estimate of $\Phi(t_0)$ which is important in the proof of Theorem 1.

Remark 2. It is clear that it suffices to prove Theorem 3 assuming that p is a step-function taking finitely many values.

The bridge from Theorem 3 to Theorem 1 is given by the following result of F. Norstad [8].

Theorem A. *Let* u *be subharmonic in an annulus* $\{R_1 < |z| < R_2\}$. *If* λ *is given,* $0 < \lambda < 1$, *and*

$$u(-r) \leq \cos \pi\lambda u(r), \quad R_1 < r < R_2,$$

then $L_\lambda(r) = L_\lambda(r, u) = \int_{-\pi}^{\pi} u(re^{i\phi})\sin \lambda(\pi - |\phi|)d\phi$ *is convex with respect to the family* $Ar^\lambda + Br^{-\lambda}$ (A, B are constants).

The case $\lambda = \frac{1}{2}$ is due to L. Ahlfors (cf. Heins [5], Ex. 4, p. 112).

Assuming Theorem 3 and Theorem A to hold, Theorem 1 is proved in the following way. Let $R > 0$ be given. Let u^* be the associated subharmonic function which is obtained by projecting the

Riesz mass of u onto the negative real axis as in HKN [6]. Let $L_\mu(r) = L_\mu(r, u^*)$. Using Norstad's argument, we see that

$$r^2 L_\mu'' + rL_\mu' - \mu^2 L_\mu = 2\mu\{u^*(r)\cos \pi\mu - u^*(-r)\}, \quad 0 < r < R.$$

Since the function $\theta \to u^*(re^{i\theta})$, $0 \le \theta < \pi$, is decreasing and we have (1), we can deduce a differential inequality for L_μ. Applying Theorem 3, we obtain Theorem 1.

To prove Theorem 2, we replace the Denjoy integral inequality in A. Baernstein ([1], Key Inequality I) by an inequality for functions subharmonic in a disc of the same type as the one discussed in HKN [6]. Theorem 2 is a consequence of this second inequality and formula (32) in Baernstein [1]. For details, we refer to Essén ([3], §8).

Some further remarks

In order to discuss the precision of (4), we re-write this formula in the following way:

$$\text{(4')} \quad M(r) \le \text{Const.}\,(|u(0)| + M(R)\exp\{-\int_r^R(\mu + C(\mu)(\psi(t) - \mu) + \text{Const.}\,(\psi(t) - \mu)^2)dt/t\}), \quad r_0 < r < R,$$

where $C(\mu) = (\pi\mu/2)\cot(\pi\mu/2)$.

More can be said if more is known about the function ψ. As an example, we consider the following situation. Let λ and ρ be given, $0 < \lambda < \rho < 1$. Assume that ψ is a stepfunction taking only the two values λ and ρ and that (1) holds. Then

$$\text{(4'')} \quad M(r) \le \text{Const.}\,M(R)\exp\{-\lambda\int_{E(r,\,R)}dt/t - \rho\int_{F(r,\,R)}dt/t + b(\lambda)(\rho - \lambda)^2\int_{E(r,\,R)}dt/t\}, \quad 0 < r < R,$$

(cf. Essén [4]). Here $F = \{r : m(r) \le \cos \pi\rho M(r)\}$, $E = \text{Compl. } F$, $E(r, R) = E \cap (r, R)$ and $F(r, R) = F \cap (r, R)$.

Formula (4") says more than formula (4') when the range of ψ is of the form $\{\lambda, \rho\}$ and $\rho - \lambda$ is small.

Problems

1. Give other examples of a class of functions $\mathfrak{F}$ such that if $\psi \in \mathfrak{F}$

and (1) holds, something more precise than (4') will be true for the growth of $M(r)$.

2. Can (4') be improved in general?

The counterexample to the Kjellberg conjecture (2) is given by the following subharmonic function. Let ρ, η and A be given positive numbers such that $\frac{1}{2} < \rho < 1$, $\eta A < \rho$. Define

$$u(z) = \mathrm{Re}\{\int_0^\infty z(z+t)^{-1}\,\nu(t)dt/t\},$$

where $\nu(t) = t^\rho \exp\{\eta \sin(A \log t)\}$.

Problem

3. Construct a counterexample to (2) where the subharmonic function has order ρ, $0 \le \rho \le \frac{1}{2}$.

References

1. A. Baernstein II. A generalization of the cos $\pi\rho$-theorem, to appear, Trans. of the A. M. S.
2. P. D. Barry. Some theorems related to the cos $\pi\rho$-theorem, Proc. London Math. Soc. XXI, (1970), 334-60.
3. M. Essén. Lectures on the cos $\pi\lambda$-theorem, University of Kentucky (1973).
4. M. Essén. Theorems of (λ, ρ)-type on the growth of subharmonic functions, Report (1971), Department of Mathematics, Royal Institute of Technology, S-10044 Stockholm 70, Sweden.
5. M. Heins. Selected topics in the classical theory of a complex variable, Holt, Rinehart and Winston, New York (1962).
6. U. Hellsten, B. Kjellberg and F. Norstad. Subharmonic functions in a circle, Arkiv f. Matematik, 8 (1970), 185-93.
7. J. Lewis. Some theorems on the cos $\pi\lambda$-inequality, Trans. A. M. S., 167 (1972), 171-89.
8. F. Norstad. Convexity of the mean-value of certain subharmonic functions (in Swedish). Manuscript (1970).

Royal Institute of Technology,
Stockholm, Sweden.

University of Kentucky,
Lexington, U. S. A.

APPLICATIONS OF DENJOY INTEGRAL INEQUALITIES TO GROWTH PROBLEMS FOR SUBHARMONIC AND MEROMORPHIC FUNCTIONS

MATTS ESSÉN[1] and DANIEL F. SHEA

Let $f(z)$ be an entire function of order ρ, and put

$$(1) \qquad A(r) = \inf_{|z|=r} \log |f(z)|, \qquad B(r) = \sup_{|z|=r} \log |f(z)| .$$

Using difficult arguments of Phragmén-Lindelöf type, Valiron [20] and Wiman [22] proved the sharp inequality

$$(2) \qquad \limsup_{r \to \infty} \frac{A(r)}{B(r)} \geq \cos \pi\rho$$

when $0 \leq \rho \leq 1$. Some twenty years later, Denjoy [5] observed that

$$(3) \qquad \int_x^\infty \{A(r) - \cos \pi\alpha\, B(r)\} \frac{dr}{r^{1+\alpha}} \geq C(\alpha) \frac{B(x)}{x^\alpha} \quad (0 < x < \infty;\ \rho < \alpha < 1),$$

where $C(\alpha)$ is a positive constant. This elegant inequality, whose proof is completely elementary, not only gave a simple proof of (2) but also became the basis for some far-reaching extensions of (2), we mention in particular those of Kjellberg [15][16] and Barry [2][3]. Using in an essential way Denjoy's precise evaluation of $C(\alpha)$,

$$(4) \qquad C(\alpha) = \min\left\{\frac{1 - \cos \pi\alpha}{\alpha}, \frac{1 + \cos \pi\alpha}{1 - \alpha}\right\},$$

Barry found that

$$(5) \qquad \underline{\log \text{dens}}\ \{r : A(r) > \cos \pi\alpha\, B(r)\} \geq 1 - \rho/\alpha \quad (\rho < \alpha < 1).$$

This result is sharp (Hayman [13]), and contains earlier theorems of Besicovitch, Beurling and Kjellberg [14].

---

1 Research supported by the Swedish Natural Science Research Council.

Our purpose here is to point out two methods which generate general inequalities of Denjoy's type (3); as applications we deduce new results on the growth of functions meromorphic in **C** or subharmonic in $\mathbf{R}^d$, $d \geq 2$, or in half-spaces of $\mathbf{R}^d$. We give only a brief discussion here, with details to appear elsewhere.

Method I. Under mild conditions on real functions g, K and G satisfying the convolution integral inequality

$$(*) \qquad G(r) \leq \int_0^\infty g(t)K(\tfrac{r}{t})\,\frac{dt}{t} \qquad (0 < r < \infty),$$

the Denjoy-type inequality

$$(6) \qquad \int_x^\infty \{\hat{K}(\alpha)g(r) - G(r)\}\,\frac{dr}{r^{1+\alpha}} \geq C(\alpha)\,\frac{g(x)}{x^\alpha} \qquad (0 < x < \infty)$$

can be shown to hold with $C(\alpha)$ given in terms of the Mellin transform $\hat{K}(\alpha) = \int_0^\infty K(t)t^{-\alpha-1}dt$ by

$$(7) \qquad C(\alpha) = \frac{\hat{K}(\alpha) - \hat{K}(0)}{\alpha}$$

By Barry's argument for deducing (5) from (3) and (4), we get from (6) and (7) that

$$\underline{\log \text{dens}}\ \{r : \hat{K}(\alpha)g(r) > G(r)\} \geq 1 - \rho/\alpha \qquad (\rho < \alpha < \tau_2),$$

where $\hat{K}(\alpha)$ converges on $\tau_1 < \alpha < \tau_2$, $\tau_1 < 0$, and g has order $\rho \in (\tau_1, \tau_2)$.

Method II. Let $v(z)$ be subharmonic in $\{\operatorname{Im} z > 0\}$ and such that boundary values $v(\pm r)$, $0 \leq r < \infty$, and partial derivatives $v_\theta(\pm r)$ can be defined. Introduce

$$(8) \qquad L(r) = \int_0^\pi v(re^{i\theta})h(\theta)d\theta\ ,$$

where h is a nonnegative solution of the differential inequality

$$(**) \qquad h''(\theta) + \alpha^2 h(\theta) \leq 0 \qquad (0 \leq \theta \leq \pi).$$

By subharmonicity of v, and (**), we get

$$r\frac{d}{dr}r\frac{d}{dr}L(r) \geq -\int_0^\pi v_{\theta\theta}(re^{i\theta})h(\theta)d\theta$$

$$\geq \alpha^2 L(r) - I(r)$$

where

$$I(r) = h'(0)v(r) - h'(\pi)v(-r) + h(\pi)v_\theta(-r) - h(0)v_\theta(r).$$

From this we deduce an inequality of Denjoy type,

$$\int_x^\infty I(r)r^{-\alpha-1}dr \geq x^{-\alpha}\{\alpha L(x) + xL'(x)\} \qquad (0 < x < \infty). \tag{9}$$

In applications, (9) is simplified by choosing h so that one of the constants $h(0)$, $h(\pi)$, $h'(0)$, $h'(\pi)$ in $I(r)$ vanishes.

For example, if $u(z)$ is subharmonic in **C** with order $\rho < 1$, and we put

$$v(re^{i\theta}) = \int_0^\theta u^*(re^{i\phi})d\phi$$

where u^* is the associated subharmonic function whose Riesz mass has been projected onto the negative real axis, then the choice $h(\theta)=\cos(\pi-\theta)\alpha$ in (9) yields

$$\int_x^\infty \{u^*(-r) - \cos\pi\alpha\, u^*(r)\}\frac{dr}{r^{1+\alpha}} \geq \frac{\alpha L(x) + xL'(x)}{x^\alpha},$$

from which Barry's density theorem (5) can be deduced.

The choice $h(\theta) = \sin(\pi - \theta)\alpha$, $0 < \alpha < 1$, with $v = u^*$ as above satisfying

$$v(-r) \leq \cos\pi\alpha\, v(r) \qquad (0 < r < \infty),$$

has been studied by Norstad [18], who establishes a convexity property for the mean (8) in this case.

As examples of new results that can be derived from (6) and (9), we discuss two problems concerning the growth of (1) subharmonic functions in $\mathbf{R}^d$, (2) meromorphic functions in **C**.

Application 1. Let $u(x) = u(x_1, \ldots, x_d)$ be subharmonic in R^d, and consider

$$B(r, u) = \max_{|x|=r} u(x), \quad T(r, u) = \sigma_d^{-1} \int_{|x|=1} u(rx)^+ d\sigma(x)$$

where $d\sigma$ denotes $(d - 1)$-dimensional surface area on $|x| = 1$ and $\sigma_d = \int_{|x|=1} d\sigma$. We assume $B(r, u)$ is unbounded and $u(x)$ has finite order $\rho = \lim\sup \log B(r, u)/\log r$. Govorov [12], when $d = 2$, and Dahlberg [4], when $d \geq 3$, found the sharp bound

(10) $$\limsup_{r \to \infty} \frac{T(r, u)}{B(r, u)} \geq C(\rho, d).$$

Here for each $\rho \in [0, \infty)$ and $d \geq 2$, $C(\rho, d) = \lim T(r, U)/B(r, U)$ for a certain extremal function $U(x)$ positive and harmonic in the cone $D_a = \{x : x_1 > a|x|\}$, vanishing on the boundary and $\equiv 0$ elsewhere in $\mathbf{R}^d$; $U(x) = U(x; \rho, d)$ and $a = a(\rho, d)$ are given in terms of Gegenbauer functions [4] when $d \geq 3$, with $U(x; \rho, 2) = |x|^\rho \cos \theta\rho$ in D_a and $a(\rho, 2) = \cos(\pi/2\rho)$ if $\rho \geq \frac{1}{2}$, $= -1$ if $\rho < \frac{1}{2}$.

We seek an estimate on the size of the set

$$G_\alpha = \{r : T(r, u) > C(\alpha, d)B(r, u)\}$$

for $\alpha > \rho$. Using Method I described above, together with a convolution inequality relating $T(r, u)$ and $B(r, u)$, we deduce

Theorem 1. If $\rho < \alpha < \infty$,

log dens $G_\alpha \geq 1 - \rho/\alpha$.

The convolution inequality just referred to is

(11) $$B(r, u) \leq \int_0^\infty T(t, u) K_a\left(\frac{r}{t}\right) \frac{dt}{t}$$

where K_a is a positive kernel given in terms of the Neumann function for D_a, $a = a(\alpha, d)$; see [4], pp. 304, 295, and [19], [10]. Inequality (11) was first worked out by Petrenko [19], for $d = 2$.

According to Method I, (11) implies

$$\int_x^\infty \{\hat{K}_a(\alpha)T(r, u) - B(r, u)\} \frac{dr}{r^{1+\alpha}} \geq \frac{\hat{K}_a(\alpha) - \hat{K}_a(0)}{\alpha} \frac{T(x, u)}{x^\alpha}$$

for $x > 0$; a straightforward application of Barry's method [2], used to deduce (5) from (3), now yields Theorem 1.

Essén and Lewis had previously proved Theorem 1 in the case $d = 2$ by a different method; see [9].

Inequality (10) can be strengthened by replacing ρ by the lower order μ [19] [4], or even by the 'lower Pólya peak index'

$$\mu_* = \inf\{\gamma : \liminf_{r,\sigma\to\infty} \frac{B(\sigma r, u)}{\sigma^\gamma B(r, u)} = 0\}$$

introduced in [6]. This follows easily from a generalization of (11).

$$B(r, u) \le \int_0^R T(t, u) K_a(\tfrac{r}{t}) \frac{dt}{t} + CT(6R)(\tfrac{r}{R})^\tau \quad (0 < r < R/2) \tag{12}$$

where C, τ are constants, $\tau > \alpha$; see [4] [10] [19]. Using (12), the proof of Theorem 1 can be modified to give a Denjoy-Kjellberg type inequality

$$\int_x^y \{\hat{K}_a(\alpha)T(r, u) - B(r, u)\} \frac{dr}{r^{1+\alpha}} \ge \frac{\hat{K}_a(\alpha) - \hat{K}_a(0)}{\alpha} \frac{T(x, u)}{x^\alpha} - C\frac{T(6y, u)}{y^\alpha}$$

when $0 < x < y < \infty$. From this we deduce

$$\overline{\log \text{dens}}\, G_\alpha \ge 1 - \mu/\alpha \qquad (\mu < \alpha < \infty), \tag{13}$$

$$\overline{\text{strong log dens}}\, G_\alpha \ge 1 - \mu_*/\alpha \qquad (\mu_* < \alpha < \infty), \tag{14}$$

and

$$\underline{\text{strong log dens}}\, G_\alpha \ge 1 - \rho_*/\alpha \qquad (\rho_* < \alpha < \infty) \tag{15}$$

where ρ_* is defined analogously to μ_* [6].

Application 2. Let $f(z)$ be meromorphic in $\mathbf{C}$, of finite order ρ. We wish to compare the growth of $A(r)$ defined in (1) with that of the Nevanlinna characteristic $T(r) = T(r, f)$. When $f(z)$ is entire, or more generally when

$$\delta(\infty) = 1 - \limsup_{r\to\infty} \frac{N(r, \infty)}{T(r)}$$

is maximal, $\delta(\infty) = 1$, we have

$$\limsup_{r\to\infty} \frac{A(r)}{T(r)} \ge \begin{cases} \pi\rho \cot \pi\rho & (0 \le \rho < \frac{1}{2}) \\ \pi\rho \cos \pi\rho & (\frac{1}{2} \le \rho < 1). \end{cases} \tag{16}$$

For $\rho < \frac{1}{2}$ this is an old result of Valiron [21], cf. [7]. The second part is a recent, as yet unpublished result of Edrei and Fuchs.

For meromorphic functions with arbitrary $\delta(\infty) \in [0, 1]$, Petrenko [19] has proved

$$(17) \qquad \limsup_{r\to\infty} \frac{A(r)}{T(r)} \geq -\pi\rho \qquad (\tfrac{1}{2} \leq \rho < \infty).$$

When $f(z)$ has $\rho < \frac{1}{2}$ and $\delta(\infty) > 1 - \cos \pi\rho$, the first part of (16) has been generalized to

$$(16a) \qquad \limsup \frac{A(r)}{T(r)} \geq \frac{\pi\rho}{\sin \pi\rho} \{\cos \pi\rho - 1 + \delta(\infty)\}$$

by Goldberg and Ostrovskii [11].

Our Theorem 2 yields the analogous extension of the second part of (16):

$$(16b) \qquad \limsup \frac{A(r)}{T(r)} \geq \pi\rho \{\sqrt{\delta(\infty)[2 - \delta(\infty)]} \cos \pi\rho - [1 - \delta(\infty)]\sin \pi\rho\}$$

for $f(z)$ with

$$\rho \in [\tfrac{1}{2}, 1) \text{ and } \delta(\infty) > 1 - \sin \pi\rho .$$

Equality can hold in any of (16), (16a), (16b), (17) for suitable $f(z)$.

Theorem 2. Denote the right side of (16b) by $\Gamma(\rho, \delta(\infty))$, and put

$$G_\alpha = \{r : A(r) > \Gamma(\alpha, \delta(\infty))T(r)\}.$$

For $f(z)$ of order $\rho < 1$, and α such that

$$(18) \qquad \rho < \alpha, \ \tfrac{1}{2} < \alpha < 1, \ \delta(\infty) > 1 - \sin \pi\alpha,$$

we have

$$(19) \qquad \text{log dens } G_\alpha \geq 1 - \rho/\alpha.$$

Analogues of (13)-(15), for which ρ is replaced by μ, μ_* or ρ_* in (18) and (19), hold also.

Inequalities (16), (16a), (17) are known to hold with ρ replaced by μ, via methods developed by Kjellberg, Edrei and Ostrovskii.

An interesting open question concerns to what extent Petrenko's inequality (17) is sharp when $\delta(\infty)$ is known, say for entire functions. The Edrei-Fuchs inequality (16) shows that in this case (17) can be improved for $\frac{1}{2} \leq \rho < 1$; for $\rho \geq 1$ there exist entire $f(z)$ for which

$$\limsup \frac{A(r)}{T(r)} = -\pi\rho\psi(\rho)$$

where

$$\psi(\rho) = \begin{cases} (1 + |\sin \pi\rho|)^{-1} & (1 \leq \rho \leq \frac{3}{2}) \\ \frac{1}{2} & (\frac{3}{2} < \rho < \infty), \end{cases}$$

so that (17) is not very far from the correct bound. (When $\rho > \frac{3}{2}$, such $f(z)$ are given by

$$f(z) = \lim_{R \to \infty} \int_{C(R)} \frac{\exp(-\zeta^\rho)}{\zeta - z} d\zeta$$

where $C(R)$ is the arc $|z| = R$, $|\arg z| \leq 3\pi/2\rho$; when $\rho \leq \frac{3}{2}$ it is sufficient to consider Lindelöf functions [17, p. 18].) For an extension of (17) which is sharp in another direction, see [10, p. 24]; cf. [1, p. 419].

Using appropriate machinery, (16b) can be proved using either Method I or II. But we are able to establish Theorem 2 in its full strength only via II; we give a sketch of the argument here.

Let α satisfy (18), and determine $\beta \in (0, \pi/2\alpha]$ by $\cos \beta\alpha = 1 - \delta(\infty)$. Put $\gamma = 1 - \beta/\pi$ and $\lambda = \gamma\alpha$. We apply Method II with α replaced by λ, $h(\theta) = \sin \theta\lambda$ and

$$v(z) = T^*(e^{i\beta} z^\gamma)$$

where

$$T^*(re^{i\theta}) = \sup_{|E|=2\theta} \frac{1}{2\pi} \int_E \log|f(re^{i\phi})| d\phi + N(r, \infty)$$

is the subharmonic function introduced by Baernstein [1]. Then

$$L_\beta(r) = \int_0^\pi v(re^{i\theta}) \sin \theta\lambda d\theta = \gamma^{-1} \int_\beta^\pi T^*(r^\gamma e^{i\theta}) \sin(\theta - \beta)\alpha d\theta$$

can be shown to be a convex function of $\log r$, and (9) implies after a change of variable

$$(20)\qquad \int_x^\infty \{\pi\alpha T^*(re^{i\beta}) - \pi\alpha\cos(\pi-\beta)\alpha N(r,\ 0) + \sin(\pi-\beta)\alpha A(r)\}\ \frac{dr}{r^{1+\alpha}} \geq \pi\alpha x^{-\alpha} L_\beta(x^{1/\gamma}).$$

Letting $\beta \to 0$ here yields also

$$(21)\qquad \int_x^\infty \{\pi\alpha N(r,\ \infty) - \pi\alpha\cos\pi\alpha N(r,\ 0) + \sin\pi\alpha A(r)\}\ \frac{dr}{r^{1+\alpha}} \geq \pi\alpha x^{-\alpha} L_0(x).$$

By (18) and our choice of β, $\cos\pi\alpha < 0$ and $\cos(\pi - \beta)\alpha > 0$; thus we can eliminate $N(r,\ 0)$ between (20) and (21) to obtain

$$\int_x^\infty \{\pi\alpha\sec(\pi - \beta)\alpha T^*(re^{i\beta}) + [\tan(\pi - \beta)\alpha - \tan\pi\alpha]A(r) - \pi\alpha\sec\pi\alpha N(r,\ \infty)\}\ \frac{dr}{r^{1+\alpha}} \geq \pi\alpha x^{-\alpha}\{\sec(\pi - \beta)\alpha L_\beta(x^{1/\gamma}) - \sec\pi\alpha L_0(x)\}.$$

Theorem 2 can be deduced from this by arguments analogous to those used by Barry in proving (5) from Denjoy's inequality (3).

Remark. The results of this note were announced at the conference on complex analysis at Canterbury (July 1973). A. Edrei and W. Fuchs have independently obtained the result stated in (16b). Their result was also announced at this conference.

References

1. A. Baernstein. Proof of Edrei's spread conjecture, Proc. London Math. Soc., (3) 26 (1973), 418-34.
2. P. D. Barry. On a theorem of Besicovitch, Quart. J. Math., Oxford (2) 14, (1963), 293-302.
3. P. D. Barry. On a theorem of Kjellberg, Quart J. Math., Oxford (2) 15, (1964), 179-91.
4. B. Dahlberg. Mean values of subharmonic functions, Arkiv Mat., 10, (1972), 293-309.
5. A. Denjoy. Sur un théorème de Wiman, C. R. Acad. Sci. (Paris), 193, (1931), 828-30.
6. D. Drasin and D. F. Shea. Pólya peaks and the oscillation of positive functions, Proc. Amer. Math. Soc., 34, (1972), 403-11.

7. A. Edrei. The deficiencies of meromorphic functions of finite lower order, Duke Math. J., 31 (1964), 1-21.

8. A. Edrei. A local form of the Phragmen-Lindelöf indicator, Mathematika, 17, (1970), 149-72.

9. M. Essén. Lectures on the cos $\pi\lambda$ theorem, mimeographed notes, Univ. of Kentucky (1973).

10. W. H. J. Fuchs. Topics in Nevanlinna theory, Proc. NRL Conference (1970).

11. A. A. Goldberg and I. V. Ostrovskii. Some theorems on the growth of meromorphic functions, Uc. Zap. Khar'kovsk. Univ., (5), vyp. 7 (1961), 3-37 (in Russian).

12. N. V. Govorov. On Paley's problem, Funk. Anal., 3, (1969), 35-40 (in Russian).

13. W. K. Hayman. Some examples related to the cos $\pi\rho$ theorem, Math. Essays Dedicated to A. J. Macintyre (Ohio Univ. Press, 1970), 149-170).

14. B. Kjellberg. On certain integral and harmonic functions, dissertation, Uppsala, (1948).

15. B. Kjellberg. On the minimum modulus of entire functions of lower order less than one, Math. Scand., 8 (1960), 189-97.

16. B. Kjellberg. A theorem on the minimum modulus of entire functions, Math. Scand., 12, (1963), 5-11.

17. R. Nevanlinna. Le théorème de Picard-Borel et la théorie des fonctions méromorphes, Paris (1929).

18. F. Norstad. Konvexitet hos medelvärdet av vissa subharmoniska funktioner, unpublished manuscript (1970).

19. V. P. Petrenko. The growth of meromorphic functions of finite lower order, Izv. Ak. Nauk U.S.S.R., 33, (1969), 414-54 (in Russian).

20. G. Valiron. Sur les fonctions entières d'ordre nul et d'ordre fini et en particulier les fonctions à correspondance régulière, Ann. Fac. Sci. Univ., Toulouse, (3) 5, (1949), 117-257.

21. G. Valiron. Sur le minimum de module des fonctions entières d'ordre inférieur à un, Mathematica (Cluj) 11, (1935), 264-9.

22. A. Wiman. Über eine Eigenschaft der ganzen Funktionen von der Höhe Null, Math. Ann., 76, (1915), 197-211.

Royal Institute of Technology,
Stockholm, Sweden

University of Wisconsin,
Madison, U. S. A.

A THEOREM ON $\min_{|z|}\log|f(z)|/T(r, f)$

W. H. J. FUCHS

The following theorem was proved independently (and with very similar proofs) by Matts Essén and Daniel F. Shea and by Albert Edrei and W. H. J. Fuchs.

Theorem. Let $f(z)$ be a meromorphic function of lower order μ, $\frac{1}{2} < \mu < 1$. Write

$$m_0(r) = \inf_{|z|=r} \log|f(z)|, \qquad \delta = \delta(\infty, f), \quad T(r) = T(r, f).$$

Then

$$\limsup m_0(r)/T(r) \geq -\pi\mu b(\mu), \tag{1}$$

where

$$b(\mu) = 1, \qquad \text{if } \delta < 1 - \sin \pi\mu, \tag{2}$$

$$b(\mu) = (1 - \delta)\sin \pi\mu - [\delta(2 - \delta)]^{\frac{1}{2}} \cos \pi\mu, \quad \text{if } \delta \geq 1 - \sin \pi\mu. \tag{3}$$

Remarks. 1. For $\mu < \frac{1}{2}$ and $\delta > 1 - \cos \pi\mu$ A. Edrei [2] and I. V. Ostrovskii [4] have shown that (1) holds with $b(\mu) = (-\cos \pi\mu + 1 - \delta)/\sin \pi\mu$. For $\mu \geq 1$ (1) continues to hold with $b(\mu) = 1$. This follows from Petrenko's generalisation of Paley's conjecture [5] applied to $1/f(z)$. It will also follow from our proof of the theorem.

2. The values of $b(\mu)$ are best possible for $0 < \mu < 1$. Whether the bound can be improved for functions with a given value of δ and $\mu > 1$ is an open question, it is best possible for $\delta = 0$, $\mu > 1$.

The proof of the Theorem uses the function

$$T^*(re^{i\theta}) = m^*(re^{i\theta}) + N(r, f),$$

where

$$m^*(re^{i\theta}) = \sup \int_E \log|f(re^{i\phi})|d\phi ,$$

the supremum taken over all measurable subsets of $[0, 2\pi)$ of Lebesgue measure 2θ.

This function was introduced by A. Baernstein [1]. He showed that it has the following properties

(4) $T^*(re^{i\theta})$ is subharmonic in $0 < r < \infty$, $0 < \theta < \pi$.

(5) If $f(0) = 1$, it has boundary values

$$T^*(r) = N(r, f), \qquad (r > 0)$$
$$T^*(-r) = N(r, 1/f).$$

(6) The derivative $\frac{\partial T^*}{\partial \theta}(re^{i\theta})$ has boundary values

$$\frac{\partial T^*}{\partial \theta}(r) = -\frac{1}{\pi}\log M(r) , \qquad (r > 0)$$
$$\frac{\partial T^*}{\partial \theta}(-r) = \frac{1}{\pi}\log m_0(r) ,$$

(7) $\sup_{0< \theta< \pi} T^*(re^{i\theta}) = T(r).$

We also need

(8) Given $a > 0$ there is an R, $a \leq R < 2a$ and an absolute constant K such that

$$\left|\frac{\partial T^*}{\partial \theta}(Re^{i\theta})\right| < KT(2R)/R \qquad (0 < \theta < \pi).$$

Proof of the Theorem. Without loss of generality we may assume $f(0) = 1$. Apply Green's formula

$$\iint_D(u\Delta v - v\Delta u)dA = \int_{\partial D}(u\frac{\partial v}{\partial n} - v\frac{\partial u}{\partial n})ds$$

with

$$D : S < r < R, \qquad 0 \leq \beta < \theta < \pi ,$$

$$u = r^{-\mu} \sin \mu(\theta - \beta),$$

$$v = T^*(re^{i\theta}).$$

This gives

$$(9) \qquad 0 \leq \sin \mu(\pi-\beta) \int_S^R r^{-\mu-1} \log m_0(r) dr - \pi\mu \cos \mu(\pi-\beta) \int_S^R r^{-\mu-1} N(r, 1/f) dr$$
$$+ \pi\mu \int_S^R r^{-\mu-1} T^*(re^{i\beta}) dr + E,$$

where

$$|E| < K\{T(2R)R^{-\mu} + T(2S)S^{-\mu}\},$$

if R and S are suitably chosen, so that the estimate (8) can be used. The choice

$$\mu(\pi - \beta) = \pi/2, \text{ i.e. } \beta = \pi - (\pi/2\mu)$$

in (9) together with (7) gives

$$(10) \qquad 0 \leq \int_S^R \{\log m_0(r) + \pi\mu T(r)\} r^{-\mu-1} dr + E.$$

Standard arguments [3] show that for arbitrarily large S and R E is very small compared to $\int_S^R T(r) r^{-\mu-1} dr$. The assertion of the Theorem with $b(\mu)$ given by (2) now follows easily from (10) for all $\mu > \frac{1}{2}$.

To obtain the value (3) for $b(\mu)$ if $\delta > 1 - \sin \pi\mu$, put $\beta = 0$ in (9) and remember (5):

$$(11) \qquad 0 \leq \sin \pi\mu \int_S^R \log m_0(r) r^{-\mu-1} dr - \pi\mu \cos \pi\mu \int_S^R N(r, 1/f) r^{-\mu-1} dr$$
$$+ \pi\mu \int_S^R N(r, f) r^{-\mu-1} dr + E.$$

Eliminate the integral involving $N(r, 1/f)$ between (9) and (11) and use the estimates (7) and

$$N(r, f) < (1 - \delta + \eta) T(r) \qquad (r > r_0(\eta);\ \eta > 0).$$

This leads to

$$0 \leq \int_S^R \{\log m_0(r) + \pi\mu \frac{(1-\delta+\eta)\cos \mu(\pi-\beta)-\cos \pi\mu}{\sin \mu\beta} T(r)\} r^{-\mu-1} dr + E.$$

Adjusting β so that the factor of $T(r)$ is as small as possible yields

$$0 \leq \int_S^R \{\log m_0(r) + (\pi\mu b(\mu) + \eta_1) T(r)\} r^{-\mu-1} dr + E \quad (S > S_0(\eta_1)),$$

where $b(\mu)$ is given by (3). From this inequality the assertion of the theorem is easily deduced.

References

1. A. Baernstein. Proof of Edrei's spread conjecture, Proc. London Math. Soc., (3) 26 (1973), 418-34.
2. A. Edrei. A local form of the Phragmén-Lindelöf indicator, Mathematika, 17 (1970), 149-72.
3. W. H. J. Fuchs. Topics in Nevanlinna theory, Proc. NRL Conference on Classical Function Theory, Washington (1970), p. 30.
4. I. V. Ostrovskii. Dokl. Ak. N. U. S. S. R., 150 (1963), 32-5 (in Russian).
5. V. P. Petrenko. Izv. Ak. N. U. S. S. R., 33 (1969), 414-54 (in Russian).

Cornell University,
Ithaca, U. S. A.

THE L^p-INTEGRABILITY OF THE PARTIAL DERIVATIVES OF A QUASICONFORMAL MAPPING

F. W. GEHRING

1. Introduction

Suppose that D is a domain in euclidean n-space R^n, $n \geq 2$, and that $f : D \to R^n$ is a homeomorphism. For each $x \in D$ we set

$$L_f(x) = \limsup_{y \to x} \frac{|f(y) - f(x)|}{|y - x|},$$

$$J_f(x) = \limsup_{r \to 0} \frac{m(f(B(x, r)))}{m(B(x, r))},$$

where $B(x, r)$ denotes the open n-ball of radius r about x and m denotes Lebesgue measure in R^n. We call $L_f(x)$ and $J_f(x)$, respectively, the maximum stretching and generalized Jacobian for the homeomorphism f at the point x. These functions are nonnegative and measurable in D, and

$$J_f(x) \leq L_f(x)^n$$

for each $x \in D$. Moreover Lebesgue's theorem implies that

$$\int_E J_f dm \leq m(f(E)) < \infty$$

for each compact $E \subset D$, and hence that J_f is locally L^1-integrable in D.

Suppose next that the homeomorphism f is K-quasiconformal in D. Then

$$L_f(x)^n \leq K J_f(x)$$

a. e. in D, and thus L_f is locally L^n-integrable in D. Bojarski has shown that a little more is true in the case where $n = 2$, namely that

L_f is locally L^p-integrable in D for $p \in [2, 2+c)$, where c is a positive constant which depends only on K. Bojarski's proof consists of applying the Caldéron-Zygmund inequality to the Hilbert transform which relates the complex derivatives of a normalized plane quasiconformal mapping. Unfortunately this elegant two-dimensional argument does not suggest what the situation is when $n > 2$.

In the present talk we give a new and quite elementary proof for the Bojarski theorem which is valid for $n \geq 2$. More precisely, we show that L_f is locally L^p-integrable in D for $p \in [n, n+c)$, where c is a positive constant which depends only on K and n. The argument depends upon an inequality relating the L^1- and L^n-means of L_f over small n-cubes, and upon a lemma which derives the integrability from this inequality. We conclude with three applications, one of which yields a new characterization for quasiconformal mappings.

University of Michigan,
Ann Arbor, U. S. A.

A HILBERT SPACE METHOD IN THE THEORY OF SCHLICHT FUNCTIONS

H. GRUNSKY and J. JENKINS

The method will be described for the class Σ of functions g meromorphic and schlicht for $|z| > 1$ with normalisation

$$g(z) = z + b_0 + b_1 z^{-1} + \dots .$$

Being concerned with extremal problems, it is enough to consider a dense subclass $\Sigma_0 \subset \Sigma$. Let Σ be the set of those functions $g \in \Sigma$ for which the image domain D is a slit domain with piecewise analytic slit d.

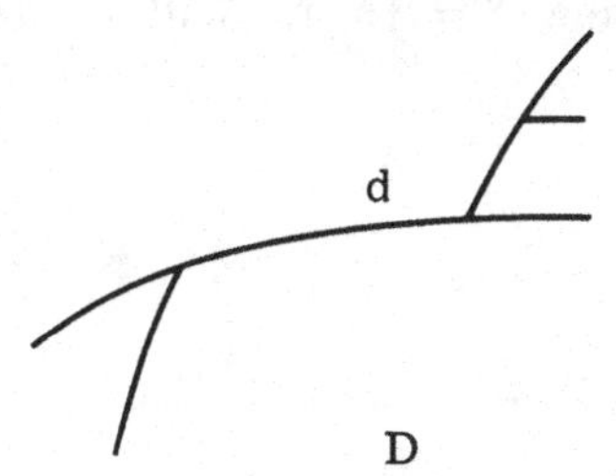

d means a pointset, the impression of ∂D which is the set of prime ends of D.

Let $\nu : d \to \mathbf{C}$ be holomorphic on d, $\infty \neq w \in D$, $\alpha \in d$. With fixed branch $\log(\alpha - w)$ is holomorphic on d and

$$(1) \qquad \int_{\partial D} \log(\alpha - w) d\nu(\alpha) = 0.$$

If, for ν, we admit functions holomorphic only in a (two-fold connected) neighbourhood of d, d excepted, with sufficiently decent behaviour on ∂D, this relation is easily shown to be characteristic for functions ν holomorphic on d.

Pull back (1) to the original plane. Let $\mathfrak{U}$ denote the unit circle described negatively. Then

$$\mu = \nu \circ g : \mathfrak{U} \to \mathbf{C}$$

is piecewise analytic, and, among all these functions, it has the characteristic property of identifying (at least) all those pairs of points on $\mathfrak{U}$ which are identified by g, i. e. it is an identifying function for g. Conversely, to every function $\mu : \mathfrak{U} \to \mathbf{C}$ which, with finitely many exceptions, identifies pairs of points of $\mathfrak{U}$, there is at least one function $g \in \Sigma_0$.

μ may be decomposed:

$$\mu = \mu_i + \mu_e,$$

where μ_i may be extended continuously to $|z| < 1$ as an analytic function, μ_e to $|z| > 1$ (interior and exterior functions). This is true for any sufficiently smooth function

$$\phi : \mathfrak{U} \to \mathbf{C} .$$

Let $B = \{\phi\}$, $I = \{\phi_i\}$ and $\mathcal{E} = \{\phi_e\}$. With an obvious notation

$$B = I + \mathcal{E} .$$

Without restriction

$$\phi_i(0) = \phi_e(\infty) = 0 .$$

Pulling back (1) we get

$$\text{(2)} \qquad \frac{1}{2\pi i} \int_{\mathfrak{U}} \mathcal{G}(z, \zeta) d\mu_i(\zeta) = \mu_e(\zeta)$$

with

$$\mathcal{G}(z, \zeta) = \log[\frac{g(z) - g(\zeta)}{z - \zeta}] \qquad \text{(Schiffer's kernel).}$$

(Note that $\mathcal{G}(z, \zeta) = \mathcal{G}(\zeta, z)$, $\mathcal{G}(\infty, \zeta) = 0$, and that $\mathcal{G}(z, \zeta)$ is holomorphic for $|z| > 1$, $|\zeta| > 1$ if and only if g is schlicht.)

We introduce the integral transformations:

$$\hat{\mathcal{G}}\, \phi(z) = \frac{1}{2\pi i} \int_{\mathfrak{U}} \mathcal{G}(z, \zeta) d\phi(\zeta) ,$$
$$\hat{\bar{\mathcal{G}}}\, \phi(z) = - \frac{1}{2\pi i} \int_{\mathfrak{U}} \overline{\mathcal{G}(z, \zeta)} d\phi(\zeta).$$

Applying (2) to

$$\mu(\zeta) = \log(\overline{g(\zeta)} - \overline{g(s)}) - \log(-\bar{s})$$

with $|s| > 1$, we find

$$\frac{1}{2\pi i} \int_{\mathfrak{U}} \mathcal{G}(z, \zeta) d_{\zeta} \overline{\mathcal{G}(\zeta, s)} = \log(1 - z^{-1}\bar{s}^{-1})$$

which means

(3a, b) $\hat{\mathcal{G}}\, \hat{\bar{\mathcal{G}}} = \mathrm{id}$ in $\mathcal{E}$, $\quad \hat{\bar{\mathcal{G}}}\, \hat{\mathcal{G}} = \mathrm{id}$ in I.

With

$$J = \hat{\mathcal{G}} + \hat{\bar{\mathcal{G}}}$$

and

$$\mu^* = i(\mu_i - \mu_e)$$

we obtain

$$J\mu = \mu, \qquad J\mu^* = -\mu^* .$$

We make B an inner product space and, by completion, a Hilbert space introducing as inner product for $\phi, \psi \in B$,

$$(\phi, \psi) = \frac{1}{2\pi} \int_{\mathfrak{U}} \phi(\zeta) d\psi^*(\zeta).$$

It is easily checked that the properties of an inner product hold. Also

$$(\phi_i, \psi_e) = 0 \quad \text{if} \quad \phi_i \in I, \ \psi_e \in \mathcal{E}$$

and so

$$B = I \oplus \mathcal{E} .$$

It is easily proved that $\hat{\mathcal{G}}$ and $\hat{\bar{\mathcal{G}}}$ are adjoint and unitary and that J is self-adjoint, involutary and unitary with eigenvalues ±1. The eigenfunctions with eigenvalue 1 are the identifying functions for g.

By Schwarz's inequality

$$(4) \qquad |(\hat{\mathcal{G}}\,\phi_i,\ \phi_e)|^2 \le (\phi_i,\ \phi_i).\,(\phi_e,\ \phi_e).$$

If ϕ_e is such that for $|\zeta| = 1$,

$$\phi_e(\zeta) = \overline{\phi_i(\zeta)} \underset{\text{def}}{=} \mathring{\bar{\phi}}_i(\zeta)$$

so that $\phi(\zeta) \in \mathbf{R}$, then

$$|(\hat{\mathcal{G}}\,\phi_i,\ \mathring{\bar{\phi}}_i)| \le (\phi_i,\ \phi_i).$$

Evaluation by the power series coefficients gives the inequalities proved by the first author in 1939. Equality holds, ϕ_i given, if g is such that $\phi_i + \mathring{\bar{\phi}}_i$ is an identifying function for g.

Let in (4),

$$\phi_e = \eta\,\hat{\mathcal{G}}_0\phi_i \qquad (|\eta| = 1),$$

$\mathcal{G}_0$ belonging to

$$g_0(\zeta) = \zeta + \varepsilon\zeta^{-1} \qquad (|\varepsilon| = 1).$$

Then we get the following coefficient inequalities:

$$(5) \qquad \Big|\sum_{j,k=1}^{\infty} C_{j,k}\,\bar{\gamma}_j\gamma_k\Big| \le \sum_{k=1}^{\infty} |\gamma_k|^2$$

with

$$C_{j,k} = \tfrac{1}{2}(\eta\varepsilon^k c_{j,k} + \bar{\eta}\,\bar{\varepsilon}^j\bar{c}_{j,k})$$

if

$$\mathcal{G}(z,\ \zeta) = -\sum_{j,k=1}^{\infty} \frac{c_{j,k}}{\sqrt{jk}}\,z^{-j}\zeta^{-k},$$

$$\phi_i(\zeta) = \sum_{\nu=1}^{\infty} \frac{\gamma_\nu}{\sqrt{\nu}}\,\zeta^{\nu}.$$

(Note: The left hand side in (5) is Hermitian, not quadratic as in the former inequalities.) In (5) we have equality at least if $g = g_0$.

If we choose in (4)

$$\phi_i(\zeta) = \sum_{\nu=1}^{n} \lambda_\nu \log(1 - t_\nu^{-1}\zeta) \qquad (\lambda_\nu \in \mathbf{C}, \ |t_\nu| > 1)$$

we obtain a set of inequalities proved by Golusin in 1947 by a variational method.

All eigenfunctions of J being known it is possible to represent the kernel $\mathcal{G}(z, \zeta)$, and also g. If there is an identifying function μ_0 for g characterising g unambiguously (e.g. if the slit d is a simple arc), then the powers μ_0^k for $k \in \mathbf{N}$ are a basis for the eigenfunctions with eigenvalue +1. Let μ_k be an orthonormalised basis. Then

$$g(z) = z + b_0 + 2 \sum_k \overline{\mu_{ki}(0)}\, \mu_{ke}(z).$$

Math. Inst. der Universität,
Würzburg, Germany.

Washington University,
St. Louis, U. S. A.

AN EXTREMAL PROBLEM CONCERNING ENTIRE FUNCTIONS WITH RADIALLY DISTRIBUTED ZEROS

SIMON HELLERSTEIN and DANIEL F. SHEA

According to a result of Edrei, Fuchs and Hellerstein [4], an entire function of finite order > 2, all of whose zeros are real, has zero as a Nevanlinna deficient value. It has been conjectured ([8], Problem 1.13) that for such functions the deficiency must tend to 1 as the order increases to ∞. We prove this statement, obtain sharp lower bounds for the deficiency, and consider some related problems.

We use the standard notation and terminology of Nevanlinna theory, as given in [7]. The deficiency of zero for an entire function $f(z)$ is

$$\delta(0,\ f) = 1 - \limsup_{r\to\infty} \frac{N(r,\ 0)}{T(r,\ f)}\ .$$

If $f(z)$ has order $\lambda < \infty$, then $f(z) = z^m g(z) \exp P(z)$ where $P(z)$ is a polynomial of degree p and $g(z)$ is a canonical product of genus k. The genus of $f(z)$ is $q = \max(p,\ k)$, and then $q \le \lambda \le q + 1$ (see e.g. [7], pp. 21-9).

The result of Edrei, Fuchs and Hellerstein mentioned above can be stated more precisely as

$$\delta(0,\ f) \ge A \quad (A = \text{absolute constant} > 0) \tag{1}$$

when $f(z)$ has only real zeros and genus q, $2 \le q < \infty$. Examples given in [4] show that when $q \le 1$, $\delta(0,\ f) = 0$ is possible for such $f(z)$ (even if $\lambda = 2$).

Similar problems have been considered by Gol'dberg [5] [6], Ostrovskii [13] and Hellerstein [9].

We shall show that (1) can be improved to a sharp bound, given implicitly in terms of the Weierstrass primary factor

$$E(z, k) = (1 - z)\exp\left(\sum_{j=1}^{k} \frac{z^j}{j}\right).$$

For any $q \geq 0$, put

$$(2) \qquad A_q = 1 - \sup_{r>0} \frac{\overset{+}{\log} r}{T(r, E(z, [q/2]))}$$

where $[\cdot]$ denotes the integral part of $\cdot$. Notice that $A_0 = A_1 = 0$.

Theorem 1. *If* $f(z)$ *is entire, of genus* q, *with real zeros only, then*

$$(3) \qquad \delta(0, f) \geq A_q$$

where

$$A_q \geq A_2 \quad (q \geq 2), \quad \frac{1}{2} > A_2 > \frac{1}{2\pi}$$

$$A_q = 1 - \frac{\pi^2/e + o(1)}{\log q} \quad (q \to \infty).$$

For every integer $q \geq 0$ *and* $\lambda \in [q, q+1]$, *there exist* $f(z)$ *of order* λ *satisfying the above hypotheses, for which equality holds in (3).*

Our methods also give

Theorem 2. *Let* $f(z)$ *be an entire function, all of whose zeros lie on the radii*

$$re^{i\theta_1}, \ldots, re^{i\theta_m} \quad (0 \leq r < \infty).$$

Then there exist constants $K = K(\theta_1, \ldots, \theta_m)$ *and* $B_q = B_q(\theta_1, \ldots, \theta_m)$ *such that, if the genus* q *of* $f(z)$ *is finite and greater than* K,

$$\delta(0, f) \geq B_q > 0,$$

where

$$\lim_{q \to \infty} B_q = 1.$$

Our results depend upon an extremal property of Weierstrass

products whose zeros lie on a single ray,

$$(4) \qquad g(z) = \prod_{n=1}^{\infty} E(\frac{z}{a_n}, q) \qquad (a_n > 0, \ \Sigma a_n^{-q-1} < \infty),$$

and exploit the relation

$$T(r, \hat{g}) \le T(r, g) \qquad (0 < r < \infty)$$

which holds between any two such products whenever

$$(5) \qquad N(r, 1/\hat{g}) \le N(r, 1/g) \qquad (0 < r < \infty);$$

this fact is implicit in work of Hellerstein and Williamson ([12], pp. 245, 247). If we choose $\rho_k \to \infty$ so that

$$N(\rho_k, 1/g)/T(\rho_k, g) \to 1 - \delta(0, g) \qquad (k \to \infty),$$

with ρ_{k+1}/ρ_k tending rapidly to ∞, then it is not difficult to see that there exist r_k $(\rho_{k-1} < r_k < \rho_k)$ and integers $m_k \to \infty$ such that

$$\hat{g}(z) = \prod_{k=1}^{\infty} E(\frac{z}{r_k}, q)^{m_k}$$

satisfies (5) as well as $N(\rho_k, 1/\hat{g}) = N(\rho_k, 1/g)$ for $k \ge 1$, and thus

$$\delta(0, g) \ge \delta(0, \hat{g}).$$

On the other hand since the $\rho_k \to \infty$ fast,

$$\limsup_{r\to\infty} \frac{N(r, 1/\hat{g})}{T(r, \hat{g})} = \sup_{r>0} \frac{N(r, 1/E(z, q))}{T(r, E(z, q))},$$

and thus

$$(6) \qquad \limsup_{r\to\infty} \frac{N(r, 1/g)}{T(r, g)} \le \alpha_q \equiv \sup_{r>0} \frac{\log^+ r}{T(r, E(z, q))},$$

an inequality which obviously is sharp in the class of all products (4).

To find the asymptotic behaviour of α_q for large q (and thus also that of the A_q in (2)), one shows that the sup in (6) must be attained near $r = 1 + 1/q$, and that

$$T(r,\ E(z,\ q)) \sim \frac{1}{\pi^2} \left(\frac{\log q}{q}\right) r^q \qquad (q \to \infty)$$

uniformly for r near 1.

One can ask whether (6) can be improved for $g(z)$ satisfying certain regularity properties, e.g. regular growth of order λ:

(7) $$\lambda = \lim_{r\to\infty} \frac{\log T(r,\ g)}{\log r} \quad \text{exists.}$$

(This question is implicit in the statement of Problem 1.13 in [8].) Consideration of the auxiliary functions $g(z)$ introduced above shows that our method involves comparing a given $g(z)$ with a possibly much less regular $\hat{g}(z)$, and thus (6) might be a good bound only for functions of irregular growth. In fact, given integral $q \geq 0$ and any $\lambda \in [q,\ q+1]$, it is possible to construct products (4) having regular growth in the sense of (7) for which equality holds in (6). Further, given $\varepsilon > 0$ and $\lambda \in (q,\ q+1)$, there exist products (4) having very regular growth in the sense of Valiron:

$$c < \frac{T(r,\ g)}{r^{\lambda}} < C \quad (0 < c,\ C < \infty),$$

such that

$$\limsup_{r\to\infty} \frac{N(r,\ 1/g)}{T(r,\ g)} > \alpha_q - \varepsilon .$$

(When $q = 0$, this and more had already been noted by Baernstein [1, p. 103] and Gol'dberg and Ostrovskii [6, p. 310].)

The bound α_q in (6), which is $0(1/\log q)$ when $q \to \infty$, can be decreased if lim sup is replaced by lim inf:

(8) $$\liminf_{r\to\infty} \frac{N(r,\ 1/g)}{T(r,\ g)} \leq \frac{|\sin \pi\lambda|}{q+s(\lambda)}$$

where $s(\lambda) = |\sin \pi\lambda|$ for $q \leq \lambda < q + \frac{1}{2}$, $s(\lambda) = 1$ for $q + \frac{1}{2} \leq \lambda \leq q + 1$ (Ostrovskii [13], Shea [14]; see also [10] [12] [15]).

When the limit in (6) and (8) exists, we have equality in (8) and also, when $\lambda > \frac{1}{2}$, (7) and even

$$\lim_{r\to\infty} \frac{T(\sigma r,\ g)}{T(r,\ g)} = \sigma^{\lambda} \qquad (0 < \sigma < \infty)$$

(Edrei and Fuchs [2] [3] for $\frac{1}{2} < \lambda < 1$ and $\lambda = 1, 2, \ldots$; Hellerstein, Shea and Williamson [11] for other λ).

Our methods can be extended to cover meromorphic functions having two radially distributed values. For example, we have

Theorem 3. Let $f(z)$ be meromorphic and of finite order in the plane, with zeros $\{a_n\}$ and poles $\{b_n\}$ lying in sectors

$$|\arg a_n| \le \eta, \quad |\arg b_n - \pi| \le \eta \qquad (0 \le \eta \le \tfrac{\pi}{6}).$$

Denote by q the least integer > 0 for which

$$\sum \frac{1}{|a_n|^{q+1}} + \sum \frac{1}{|b_n|^{q+1}} < \infty .$$

Then

$$\limsup_{r\to\infty} \frac{N(r, 0) + N(r, \infty)}{T(r, f)} \le [1 + \varepsilon(\eta, q)]\gamma_q \tag{9}$$

where

$$\gamma_q \equiv \sup_{r>0} \frac{2 \overset{+}{\log} r}{T(r, E(z, q)/E(-z, q))}$$

satisfies

$$\gamma_q \le \gamma_1 < 1 \quad (q \ge 1), \quad \gamma_q = \frac{2\pi^2/e + o(1)}{\log q} \qquad (q \to \infty),$$

and

$$\varepsilon(\eta, q) = C\, \eta \log(\tfrac{1}{\eta}) \{1 + q(\overset{+}{\log} q)^2\} \tag{10}$$

where C is an absolute constant.

This statement is sharp in that equality can hold in (9) for any $q \ge 0$ when $\eta = 0$; also, the dependence of ε on η and q given by (10) is essentially exact for small $\eta \cdot q$, as is shown by some examples of Gol'dberg [5, p. 391] [6, p. 346].

References

1. A. Baernstein. A nonlinear tauberian theorem in function theory and some results on tauberian oscillations, Thesis, Univ. of Wisconsin, (1968).

2. A. Edrei and W. Fuchs. Valeurs déficientes et valeurs asymptotiques des fonctions méromorphes, Comment. Math. Helv., 33 (1959), 258-95.

3. A. Edrei and W. Fuchs. Tauberian theorems for a class of meromorphic functions with negative zeros and positive poles, Proc. Conference, Erevan (1965).

4. A. Edrei, W. Fuchs and S. Hellerstein. Radial distribution and deficiencies of the values of a meromorphic function, Pacific J. Math., 11 (1961), 135-51.

5. A. A. Gol'dberg. Distribution of values of meromorphic functions with separated zeros and poles, Soviet Math., 2 (1961), 389-92.

6. A. A. Gol'dberg and I. V. Ostrovskii. The distribution of values of meromorphic functions, Izdat. Nauka, Moscow (1971).

7. W. J. Hayman. Meromorphic functions, Oxford University Press (1964).

8. W. K. Hayman. Research problems in function theory, Athlone Press, London (1967).

9. S. Hellerstein. On a class of meromorphic functions with deficient zeros and poles, Pacific J. Math., 13 (1963), 115-24.

10. S. Hellerstein and D. F. Shea. Bounds for the deficiencies of meromorphic functions of finite order, Proc. Symposia Pure Math. 11, Amer. Math. Soc., Providence (1968), 214-39.

11. S. Hellerstein, D. F. Shea and J. Williamson. A tauberian theorem characterizing the growth of a class of entire functions, Duke Math. J., 38 (1970), 489-99.

12. S. Hellerstein and J. Williamson. Entire functions with negative zeros and a problem of R. Nevanlinna, J. d'Analyse Math., 22 (1969), 233-67.

13. I. V. Ostrovskii. Some asymptotic properties of entire functions with negative zeros, Zap. Mat. Otd. Fiz.-Mat. i Kharkov Mat. Obsc., (4), 28 (1961), 23-32.

14. D. F. Shea. On the Valiron deficiencies of meromorphic functions of finite order, Trans. Amer. Math. Soc., 124 (1966), 201-27.

15. J. Williamson. Meromorphic functions with negative zeros and positive poles and a theorem of Teichmüller, Pacific J. Math., 42 (1972), 795-816.

University of Wisconsin,
Madison, U. S. A.

LOCAL BEHAVIOR OF SUBHARMONIC FUNCTIONS

A. HUBER

Report on some of the results which have been obtained by R. Gygax in his dissertation (ETH Zurich 1973):

Gygax has investigated functions of the type $e^{u(z)}$, where $u(z)$ is δ-subharmonic (i. e. representable as a difference of subharmonic functions) in a domain of the complex z-plane. The study of these functions is motivated by an important result of I. G. Reshetnjak (DAN SSSR 94 (1954), 631-4) who proved that every twodimensional manifold of bounded curvature in the sense of A. D. Alexandrov can locally be generated by a line element of the form $e^{u(z)}|dz|$, where $u(z)$ is δ-subharmonic.

Theorem 1. *Let* $u = v - w$ *be* δ-*subharmonic in the domain* Ω. *Then*

$$\lim_{r\to 0} \frac{1}{2\pi} \int_{\theta}^{2\pi} \exp u(z_0 + re^{i\phi})d\phi = \exp u(z_0)$$

for every $z_0 \in \Omega$, *where* $(v(z_0), w(z_0)) \neq (-\infty, -\infty)$.

Theorem 2. *Let* u *be* δ-*subharmonic in* Ω, $z_0 \in \Omega$. *Then*

$$\lim_{r\to 0} \left\{\frac{1}{2\pi} \int_{\theta}^{2\pi} \exp u(z_0 + re^{i\phi})d\phi\right\}^{\frac{1}{\log 1/r}} = \exp \mu(\{z_0\}),$$

where μ *denotes the measure associated with* u.

Remark. It has long been known that

$$\lim_{r\to 0} \frac{1}{\log 1/r} \int_{\theta}^{2\pi} u(z_0 + re^{i\phi})d\phi = \mu(\{z_0\}).$$

Recently (M. Arsove and A. Huber, Indiana Math. J. (1973)) it has been proved that

$$\lim_{r\to 0} \frac{u(z_0 + re^{i\phi})}{\log 1/r} = \mu(\{z_0\})$$

for all $e^{i\phi}$ on the unit circle with the exception of a set of logarithmic capacity zero.

Furthermore, Gygax considers a fixed domain Ω in the complex plane. Let H denote the set of all function δ-subharmonic in Ω which are locally representable as a difference of bounded subharmonic functions. Let

$$H_0 = \{u \in H \mid \inf_K u > 0 \text{ for every subcompact } K \text{ of } \Omega\}$$

$$H_1 = \{\exp u \mid u \in H\}$$

$$H_1' = \{\exp u_1 - \exp u_2 \mid u_1, u_2 \in H\}$$

Theorem 3. $H_0 = H_1$.

Theorem 4. $H_1' = H$.

It follows that H_1' is an algebra.

Eidg. Tech. Hochs.,
Zürich, Switzerland.

A GENERAL FORM OF THE ANNULUS THEOREM

JAMES A. JENKINS

The material for this talk was drawn from my joint work with Nobuyuki Suita.

The term 'annulus theorem' is used to refer to the result that if the circular ring $D : r < |z| < R$ is mapped by a regular function f into the circular ring $\Delta : r' < |w| < R'$ in such a way that concentric circles have images of index n, then $(R/r)^{|n|} \leq (R'/r')$ with equality only for $f(D)$ an unbranched unbordered $|n|$-fold covering of Δ. The result evidently has an immediate extension to any non-degenerate doubly-connected domains. It was, to the best of my knowledge, first published by Schiffer [3]. Using the definition of the module of a doubly-connected domain by the method of the extremal metric the corresponding result for univalent functions is immediate and the method can be used to prove the result stated above by a technique of choosing an appropriate covering set in the Riemann image of D [1]. If one confines attention to abstractly given doubly-connected domains perhaps little more can be said but if one considers Δ as imbedded in the Riemann sphere extensions were already pointed out in [1] whereby the image of D need not be restricted to lie in Δ provided the images of level curves (corresponding to concentric circles) of D have an appropriate index behaviour near $|z| = r$ and $|z| = R$.

The most satisfactory formulation is obtained if, instead of assigning D and Δ and placing restrictions on f, we choose D and f and define Δ properly. We may take D to be an open Riemann surface with a regular partition (A, B) of its ideal boundary, A, B not void. Let $\{D_\nu\}$ be an exhaustion of D by canonical subdomains with boundaries divided into sets of contours α_ν, β_ν corresponding to A, B where the α_ν are to be sensed negatively, the β_ν positively with respect to D_ν. Let f be a regular function on D, m, n integers,

$m > n$. Let $P_m(\alpha_\nu)$ denote the set of points of the sphere for which $f(\alpha_\nu)$ has index at least m, $Q_n(\beta_\nu)$ the set for which $f(\beta_\nu)$ has index at most n. It is readily seen that $\mathrm{Cl}P_m(\alpha_\nu) \subset \mathrm{int}\,\mathrm{Cl}P_m(\alpha_\mu)$ for $\mu < \nu$ and that $\mathrm{Cl}Q_m(\beta_\nu) \subset \mathrm{int}\,\mathrm{Cl}Q_m(\beta_\mu)$ for $\mu < \nu$. If all $P_m(\alpha_\nu)$, $Q_n(\beta_\nu)$ are not void we have non-void disjoint closed sets, $P_m(A) = \bigcap_\nu \mathrm{Cl}P_m(\alpha_\nu)$, $Q_n(B) = \bigcap_\nu \mathrm{Cl}Q_n(\beta_\nu)$.

Theorem. Let $m(D)$ denote the module of the class of locally rectifiable cycles on D separating A and B. Let f be a regular function on D for which there are non-void sets $P_m(A)$, $Q_n(B)$. Let Δ denote the union of the (finite number of) domains in the complement of $P_m(A) \cup Q_n(B)$ which have points of both sets on their boundary. Let $m(\Delta)$ denote the module of the class of rectifiable cycles in Δ separating $P_m(A)$ and $Q_n(B)$. Then $(m - n)m(D) \leq m(\Delta)$.

If $m(D)$ is finite, equality can occur only if f is a $(m - n, 1)$ map of D onto Δ apart possibly from a relatively closed subset of logarithmic capacity zero of Δ. Further the indexes of $f(\alpha_\nu)$, $f(\beta_\nu)$ with respect to each point of $P_m(A)$, $Q_n(B)$ are respectively equal to m and n for each ν.

These results can be proved either by the method of the extremal metric or by the method of harmonic length.

The portion of the statement concerning the inequality will appear shortly in the Illinois Journal [2]. The equality statement will be published subsequently.

References

1. James A. Jenkins. Some results related to extremal length, Annals of Mathematics Studies, no. 30 (1953), 87-94.
2. James A. Jenkins and Nobuyuki Suita. On regular functions on Riemann surfaces, to appear in Illinois Journal of Mathematics.
3. M. Schiffer. On the modulus of doubly-connected domains, Quart. Journal of Mathematics, 17 (1946), 197-213.

Washington University,
St. Louis, U. S. A.

TWO PROBLEMS ON H^p CLASSES

J. P. KAHANE

The functions we consider are defined on the circle **T**, and they belong to $L^1(\mathbf{T})$. This Fourier expansion is

$$f \sim \sum_{-\infty}^{\infty} \hat{f}_n e_n \qquad (e_n(t) = \exp 2\pi int)$$

H^p $(1 \leq p \leq \infty)$ is the subspace of L^p consisting of functions f such that

$$f \sim \sum_{0}^{\infty} \hat{f}_n e_n \, .$$

An inner function is a function in H^∞ whose absolute value is 1 a. e. on **T**.

The first problem deals with division of functions in H^1 by inner factors. Precisely, we are looking for classes of functions $\mathcal{C}$ ('good classes') such that, whenever $f \in \mathcal{C} \cap H^1$ and $f = gu$, where u is inner and $g \in H^1$, then $g \in \mathcal{C}$. In short, which properties are preserved through division by inner factors?

The second problem deals with metric projections from L^p to H^p. When $f \in L^p$ $(1 \leq p \leq \infty)$, there exists at least one $g \in H^p$ such that $d(f, g) = d(f, H^p)$, d being the distance on L^p. For $p = 2$, g is the orthogonal projection. For $1 < p < \infty$, the uniform convexity of the space L^p implies that g is unique. For $p = 1$, the uniqueness of g was proved by Doob (1942). For $p = \infty$, g is not unique in general (nice examples by H. S. Shapiro), but g is unique if $f \in C + H^\infty$. The mapping $T_p : f \to g$ which maps L^p (resp. $C + H^\infty$ in case $p = \infty$) on H^p is the metric projection from L^p to H^p. Again, we are looking for good classes of functions (depending on p), that is, classes of $\mathcal{C}$ such that $T_p \mathcal{C} \subset \mathcal{C}$.

The first problem has been studied recently by Havin and his students (Seminar of the Leningrad University, 1971) Gurarii (ibidem, 1972), Rabindranathan (Indiana Journal, 1972), Korenbljum and Faivišeiski (Ukrainian Journal, 1972).

The second problem was introduced in 1952 by H. S. Shapiro. An old result of F. Riesz (1920) is related to the case $p = 1$. The case $p = \infty$ has been investigated by Carleson and Jacobs (Arkiv für Matematik, 1972). Carleson proved also $T_p L^q \subset L^q$ when $p < q < \infty$ (Comptes rendus, 1973). There are some unpublished or partially published results on the case $p = 1$.

Here is a summary of results.

Λ_α $(0 < \alpha < 1)$ is the usual Lipschitz class of order α.

Given p $(1 \le p \le \infty)$ and α $(\alpha > 0)$, Λ_α^p is the class of all f such that

$$E_n^{(p)}(f) = \inf \| f - \sum_{-n}^{n} a_j e_j \|_p = O(n^{-p}) \qquad (n \to \infty).$$

Then $\Lambda_\alpha = \Lambda_\alpha^\infty$ for $0 < \alpha < 1$.

Given $0 < \rho_0 \le \rho_1 \le \rho_2 \le \ldots$, $H^2(\rho)$ is the subspace of H^2 consisting for the $f \in H^2$ such that

$$\sum_0^\infty \rho_n |\hat{f}_n|^2 < \infty.$$

W^+ consists of $f \in H^1$ such that $\sum_0^\infty |\hat{f}_n| < \infty$.

C^∞, $\mathfrak{A}$, $\mathcal{T}_n$ are the classes of infinitely differentiable functions on $\mathbf{T}$, or analytic on $\mathbf{T}$, or trigonometric polynomials of degree $\le n$.

Concerning the first problem:

Λ_α $(0 < \alpha < 1)$ is a good class (Havin). More generally, Λ_α^p is a good class.

$H^2(\rho)$ is a good class (Rabindranathan, Korenbljum and Faivišeiski).

W^+ is not a good class; precisely, there exist $f \in W^+$, $g \in H^\infty$, such that $f = g \exp \frac{z-1}{z+1}$, and $g \notin W^+$ (Gurarii).

Concerning the second problem, here is the situation: we write yes if the class is good for T_p, and no if not.

class \ p	1		2		∞
L^q with $p < q < \infty$	yes	yes	yes	yes	×
L^∞ or C	no	no	no	no	×
Λ_α $(0 < \alpha < 1)$	yes	no	yes	no	yes
Λ^p_α	yes	no	yes	no	?
C^∞ or $\mathfrak{A}$	yes	no	yes	no	yes
$\mathcal{T}_n$	yes	no	yes	no	no

The proofs were given in the lecture only for positive results concerning the division problem, and concerning the second problem in case $p = 1$. They depend on the properties of the Toeplitz operator $f \to Tf = P\bar{\phi}f$, associated with a $\phi \in H^\infty$ (se suppose $|\phi| \leq 1$). For $f \in L^1$, Pf is the trigonometric series

$$\sum_0^\infty \hat{f}_n e_n$$

or (when it is a Fourier series) the function it represents. Using best approximation in L^2, an elementary proof is given of the formula

$$\sum_{N+1}^\infty |\hat{g}_n|^2 \leq \sum_{N+1}^\infty |\hat{f}_n|^2 \qquad (g = Tf)$$

which implies

$$TH^2(\rho) \subset H^2(\rho).$$

A slightly more elaborate use of best approximation gives

$$T\Lambda^p_\alpha \subset \Lambda^p_\alpha.$$

Faculte des Sciences,
Orsay, France.

APPROXIMATION ON CURVES BY LINEAR COMBINATIONS OF EXPONENTIALS*

J. KOREVAAR

Let $p_1, p_2, \ldots$ be distinct positive integers. A classical result of Müntz (1914) and Szász (1915) implies that the exponentials $e^{p_n s}$ span the space $C_0[-\infty, 0]$ (of continuous functions vanishing at $-\infty$) if and only if the series $\Sigma\, 1/p_n$ diverges. The work of Clarkson and Erdős (1943) and of L. Schwartz (1943) shows that for convergent $\Sigma\, 1/p_n$, only very special (analytic) functions can be approximated.

Instead of the half-line $[-\infty, 0]$, we consider simple curves γ in the complex plane which may extend to infinity in a westerly direction. The question is for which curves γ one has approximation theorems similar to those indicated above.

Theorem 1. _Let γ be an arbitrary twice continuously differentiable arc. When $\Sigma\, 1/p_n$ converges, the exponentials $e^{p_n \zeta}$ fail to span $C(\gamma)$. More generally, let $\{z_n\}$ be any sequence of complex numbers such that $\Sigma\, 1/|z_n|$ converges. Then there is a nonzero measure $d\mu$ on γ whose Laplace transform_

$$f(z) = \mathcal{L}\, d\mu = \int_\gamma e^{z\zeta} d\mu(\zeta)$$

vanishes at the points z_n (and at many other points).

It is likely that the smoothness condition on γ can be relaxed.

Theorem 2. _Let γ be a curve starting at 0 and possibly extending to infinity. It is required that there be a positive constant $\alpha < \frac{1}{2}\pi$ such that the angles between the chords of γ and the negative real axis do not exceed α. When $\Sigma\, 1/p_n$ diverges, the exponentials $e^{p_n \zeta}$ span_

* Work supported in part by NSF grants GP-8445 and GP-38584.

$C_0(\gamma)$. *More generally, let* $d\mu$ *be a nonzero measure on* γ *and* $f(z)$ *its Laplace transform. Choosing* $0 < \varepsilon < \frac{1}{2}\pi - \alpha$ *and* $\delta > 0$, *we let* $\{z_n\}$ *be any sequence of zeros of* $f(z)$ *lying in the angle* $|\arg z| < \frac{1}{2}\pi - \alpha - \varepsilon$ *and satisfying the separation condition* $|z_n - z_k| \geq \delta|n - k|$. *Then*

$$\Sigma \frac{1}{|z_n|} < \infty .$$

One can not allow 'verticality' in γ, and the condition $|\arg z| < \frac{1}{2}\pi - \alpha - \varepsilon$ can not be relaxed to $|\arg z| < \frac{1}{2}\pi - \alpha$. However, the separation condition is probably unnecessary.

In the case of Theorem 1, no exponential $e^{p_k \zeta}$ is in the closed span S'_k of the other exponentials $e^{p_n \zeta}$ on γ. Setting $c = \max \operatorname{Re} \zeta$ for $\zeta \in \gamma$, one has

$$d(e^{p_k \zeta}, S'_k) > e^{(c-\varepsilon)p_k} \quad \text{as } k \to \infty$$

for every number $\varepsilon > 0$.

Theorem 3. *Let* γ *be a twice continuously differentiable arc which does not contain a vertical segment (or at least not a boundary segment of the half-plane* $H_c(\operatorname{Re} \zeta < c)$, *where* c *is as above). Let* $\Sigma\, 1/p_n$ *be convergent. Then the closed span* $S = S(e^{p_n \zeta}, n = 1, 2, \ldots)$ *in* $C(\gamma)$ *consists of the continuous functions* $g(\zeta)$ *on* γ *which on* $H_c \cap \gamma$ *are given by a convergent (power) series* $\Sigma\, a_n e^{p_n \zeta}$ *involving only the exponents* p_n.

It would be important to have very precise lower bounds for $d(1, S)$ or $d(e^{p\zeta}, S)$, $p \neq p_n$. Such bounds might help settle Macintyre's Conjecture (1952): A nonzero entire function given by a gap series $\Sigma\, a_n z^{p_n}$, with $\Sigma\, 1/p_n$ convergent, can not be bounded on any curve going to infinity. There is a fair amount of literature related to this Conjecture (cf. Kővari, 1965).

Two methods have proved useful in dealing with approximation on curves. One is the method of infinite order differential operators. It requires the introduction of suitable classes of functions $\mathcal{C}\{M_n\}$ on γ, and the development of appropriate extensions of the Denjoy-Carleman theory of quasianalyticity to curves. In lectures at Imperial College, London (Spring 1971), the author developed this method to obtain Theorem 2 and a weaker version of Theorem 1. (Cf. Korevaar in Proc. Conference on Approximation Theory, Univ. of Texas, to be published by Acad. Press, 1973.) A similar early version of Theorem 1 was obtained independently by Malliavin and Siddiqi (C. R. Paris, 1971), who also used differential operators. The present version of Theorem 1, as well as Theorem 3, were obtained by the method of placing point masses on γ: If $f(z) = \mathcal{L}\, d\mu$ vanishes at $p_1, \ldots, p_k$, then

$$m_k = d\{1, S(e^{p_n \zeta}, n = 1, \ldots, k)\} \geq |f(0)| / \int_\gamma |d\mu| .$$

University of California,
San Diego, U. S. A.

TWO RESULTS ON MEANS OF HARMONIC FUNCTIONS

Ü. KURAN

1. The first result [6] is the following

Theorem 1. Let D be a domain of finite volume in R^n where $n \geq 2$. Suppose that there exists a point P_0 in D such that, for every function h harmonic in D and integrable over D, the volume mean of h over D equals $h(P_0)$. Then D is a ball (disc when $n = 2$) centred at P_0.

Epstein [1] introduced and proved this result in the case where $n = 2$ and D is simply connected. Epstein and Schiffer [2] proved it when $R^n \backslash D$ has a non-empty interior. Recently, Goldstein and Ow [4] proved Theorem 1 when $n = 2$ and the boundary ∂D of D in R^2 has at least one component which is a continuum.

To prove Theorem 1 we first note that there exists a point, P_1, say, of $R^n \backslash D$ at shortest distance from P_0. Hence the ball B of centre P_0 and radius $a = P_0P_1$ is in D. It is easy to show that our result will be proved if we proved that

$$\partial D \subset B \cup \partial B. \tag{1}$$

Suppose, by contradiction, that (1) were not true. Then there would exist an open subset, ω, say, of D outside $B \cup \partial B$ and having positive volume. Let h be the function in $R^n \backslash \{P_1\}$ given by

$$h(P) = a^{n-2} \{(P_0P)^2 - a^2\} (P_1P)^{-n} + 1.$$

It is easy to verify that h satisfies all hypotheses. Hence

$$0 = \int_D h(P)dP = \int_B h(P)dP + \int_{D\backslash B} h(P)dP$$
$$= 0 + \int_{D\backslash B} h(P)dP \geq \int_{D\backslash B} dP \geq \int_\omega dP > 0,$$

which gives the required contradiction.

2. The second result [7] shows that, for any harmonic function h and for $0 < p < 1$, the function $|h|^p$ has a subharmonic 'behaviour'. We have

Theorem 2. Let h be harmonic in an open set Ω in R^n where $n \geq 2$, and let $p > 0$. Then there exists a positive constant $C(p, n)$ depending on p and n such that

$$|h(P)|^p \leq C(p, n)\,\mathcal{A}(|h|^p, P, r) \qquad (2)$$

whenever the ball $B(P, r)$ is in Ω, where $\mathcal{A}(|h|^p, P, r)$ denotes the volume mean of $|h|^p$ in $B(P, r)$.

We note first that $C(p, n) = 1$ when $p \geq 1$, since then $|h|^p$ is subharmonic. We exclude this trivial case $p \geq 1$ from now on.

Next, we note that (when $0 < p < 1$) the inequality (2) does not hold when $\mathcal{A}$ is replaced by the peripheral mean $\mathfrak{M}$ (on the sphere). The counterexample is the function

$$h(P) = \{1 - (OP)^2\}(IP)^{-n} \qquad (OP < 1)$$

where $I = (1, 0, \ldots, 0)$. Note that $|h(O)|^p = 1$, but the peripheral mean $\mathfrak{M}(|h|^p, 0, r) \to 0$ as $r \to 1-$. (In fact the greatest harmonic minorant H of the superharmonic function $|h|^p = h^p$ is identically zero: $0 \leq H \leq h^p$ yields that H vanishes everywhere on the unit sphere except at I, and hence $H \equiv ch$, which yields $0 \leq ch^{1-p} \leq 1$ and consequently $c = 0$.)

In the case of the plane $(n = 2)$, Theorem 2 was proved by Hardy and Littlewood [5]. To prove Theorem 2 for any Euclidean space of dimension $n \geq 2$, we make use of two results. The first result is by Flett [3] and involves the norms of the gradients of h of order 1, 2, ... defined by

$$|\nabla_1 h| = \{\sum_1^n (\frac{\partial h}{\partial x_i})^2\}^{\frac{1}{2}}, \quad |\nabla_2 h| = \{\sum_1^n (\frac{\partial^2 h}{\partial x_i^2})^2 + 2\sum_{i \neq j} (\frac{\partial^2 h}{\partial x_i \partial x_j})^2\}^{\frac{1}{2}}$$

and in an obvious way for higher orders.

Theorem (Flett [3]). *Under the hypotheses of Theorem 2 and for any positive integer* m, *there exists a positive constant* $C(p, n, m)$ *such that, for all*

$$p \geq \frac{n-2}{m+n-2}, \tag{3}$$

we have

$$\{r^m |\nabla_m h(P)|\}^p \leq C(p, n, m)\, \mathfrak{A}(|h|^p, P, r). \tag{4}$$

The second result we need helps replacing (4) by (2) and, by the converse process, replaces (3) in Flett's theorem by $p > 0$.

Theorem ([7]). *Given* h *harmonic in the unit ball, with Taylor expansion*

$$h(x_1, \ldots, x_n) = a_0 + H_1 + \ldots + H_{m-1} + H_m + \ldots$$

with H_k *denoting a polynomial of degree* k *in the variables* $x_1, \ldots, x_n$, *there exist* $N = N(m, n) > 0$, *orthogonal transformations* $T_1, \ldots, T_N$ *of* R^n *and positive constants* $\lambda_1, \ldots, \lambda_N$ *(the* λ_k *depending on* m, n *only) such that*

$$\lambda_1(h \circ T_1) + \ldots + \lambda_N(h \circ T_N) = a_0 + H^*_m + H^*_{m+1} + \ldots,$$

where H^*_k $(k \geq m)$ *is a polynomial of degree* k.

References

1. B. Epstein. On the mean-value property of harmonic functions, *Proc. Amer. Math. Soc.*, 13 (1962), 830.
2. B. Epstein and M. M. Schiffer. On the mean-value property of harmonic functions, *J. Anal. Math.*, 14 (1965), 109-11.
3. T. M. Flett. Inequalities for the pth mean values of harmonic and subharmonic functions with $p \leq 1$, *Proc. London Math. Soc.*, (3) 20 (1970), 249-75.
4. M. Goldstein and W. H. Ow. On the mean-value property of harmonic functions, *Proc. Amer. Math. Soc.*, 29 (1971), 241-4.

5. G. H. Hardy and J. E. Littlewood. Some properties of conjugate functions, J. für Math., 167 (1931), 405-23.

6. Ü. Kuran. On the mean-value property of harmonic functions, Bull. London Math. Soc., 4 (1972), 311-2.

7. Ü. Kuran. Subharmonic behaviour of $|h|^p$ ($p > 0$, h harmonic), J. London Math. Soc., (1974) (in print).

University of Liverpool,
Liverpool, England.

THE FATOU LIMITS OF OUTER FUNCTIONS

A. J. LOHWATER and G. PIRANIAN

A function defined in the unit disk D is called an outer function provided it has a representation

$$f(z) = \exp\left\{\frac{1}{2\pi i}\int_{-\pi}^{\pi}\frac{e^{it}+z}{e^{it}-z}\,d\mu(t)\right\},$$

where μ is a nonincreasing, absolutely continuous function on the interval $[-\pi, \pi]$. Until recently, the assumption prevailed that the absolute continuity of μ forces reasonably smooth boundary behaviour on the corresponding outer function f; but in 1971, we constructed an outer function f whose cluster set at each point $e^{i\theta}$ consists of the closed unit disk and whose set of Fatou limits has 2-dimensional measure π. Each Fatou limit of f has modulus less than 1.

We have now constructed an outer function f with the property that on each arc of the unit circle C, each point w of modulus less than 1 occurs uncountably often as Fatou limit of f. By necessity, the set of points $e^{i\theta}$ where w serves as Fatou limit of f is a set of first category on C. This is a consequence not of the absolute continuity of μ, but of a theorem of E. F. Collingwood and of the fact that the (empty) set on C where f has a one-point cluster set is open.

Our description of f is based on the construction of the Riemann surface S onto which f maps the disk D. The surface is unbranched, so that f' has no zeros in D. Moreover, its area is finite, and consequently Fejér's theorem implies that at each point $e^{i\theta}$, the sequence of partial sums of the power series of f reflects fairly accurately the behaviour of f along the corresponding radius of D.

The construction has obvious analogues that yield holomorphic functions in D for which each point w in the finite plane (or each point in the extended plane) occurs uncountably often as Fatou limit, on each arc of the unit circle C.

Case Western Reserve University,
Cleveland, U. S. A.

University of Michigan,
Ann Arbor, U. S. A.

A PROOF OF $|a_4| \leq 4$ BY LOEWNER'S METHOD

ZEEV NEHARI*

Loewner's classical differential equation [4] provides explicit representations for the coefficients a_n of a class of univalent functions which is dense in S. The only variable element in the expression $\psi_n[K(t)]$ for a given coefficient a_n is a function $K(t)$ of modulus 1, and it may therefore be said that this reduces the verification of the Bieberbach conjecture for a specific n to a calculus problem. Loewner himself showed that $|\psi_3(K)| \leq 3$, but the functionals $\psi_n(K)$ look more and more discouraging as n increases, and the subsequent proofs of the Bieberbach conjecture for $n = 4$ [2], $n = 5$ [7], and $n = 6$ [5, 6] were based on different methods.

We wish to show that there exists the possibility of a direct attack on the problem of finding $\max|\psi_n(K)|$, and we shall illustrate the method in the case $n = 4$. While it is true that $|a_4| \leq 4$ can now be proved by an argument which is both simple and elementary [1, 3], our basic procedure is also applicable to higher coefficients. There are, of course, technical difficulties, but the case $n = 5$ may be worth a try, especially in view of the complexity of the recent proof of $|a_5| \leq 5$ by Pederson and Schiffer [7].

We introduce the abbreviations

$$A_m = \int_0^\infty K^m(t)e^{-mt}dt, \qquad B = \int_0^\infty K^2(t)e^{-2t}\int_0^t K(s)e^{-s}ds, \tag{1}$$

$$u = K(t)e^{-t}, \qquad v = \int_t^\infty K(s)e^{-s}ds, \tag{2}$$

and we note that Loewner's expression for a_4 is

$$a_4 = -8A_1^3 + 8A_1A_2 + 4B - 2A_3 . \tag{3}$$

---

* This work was supported by the National Science Foundation under NSF Grant GP23112 A3.

Our treatment of (3) is based on the easily verified identities

$$(4) \qquad -\int_0^\infty u(u + v + A_1(\sigma - 1))^2 = \frac{a_4}{2} - (\sigma + 2)\phi + (\tfrac{5}{3} - \sigma^2)A_1^3 ,$$

$$(5) \qquad \int_0^\infty e^{-t}|u + v + A_1(\sigma - 1)|^2 = \tfrac{1}{3} + \rho^2|\sigma|^2,$$

where

$$(6) \qquad \rho = |A_1| , \qquad \phi = 2A_1A_2 - A_1^3$$

and σ is an arbitrary complex parameter. It is evident from (4) and (5) that

$$(7) \qquad \left|\frac{a_4}{2} - (\sigma + 2)\phi + (\tfrac{5}{3} - \sigma^2)A_1^3\right| \le \tfrac{1}{3} + \rho^2|\sigma|^2.$$

In addition, we shall require the inequality

$$(8) \qquad |\phi|^2 \le 2\rho^2(1 - \rho^2),$$

which follows from

$$|\phi|^2 = 4\rho^2\left|\int_0^\infty u(u - v)\right|^2 \le 4\rho^2\int_0^\infty e^{-2t}\int_0^\infty |u - v|^2$$

$$= 2\rho^2[\tfrac{1}{2} - \rho^2 + \int_0^\infty |v|^2]$$

and $\int_0^\infty |v|^2 \le \tfrac{1}{2}$.

If the assertion

$$(9) \qquad |a_4| \le 4$$

were not always true, there would exist (by a trivial argument) a function of S for which $a_4 = 4$. By (7), this would imply

$$(10) \qquad \left|\tfrac{1}{3} + \tfrac{5}{3}(1 + A_1^3) - (\sigma + 2)\phi - \sigma^2A_1^3\right| \le \tfrac{1}{3} + \rho^2|\sigma|^2.$$

Our proof of (9) consists in showing that, for any A_1 such that $|A_1| < 1$ and any ϕ subject to (8), it is possible to find a complex constant σ for which (10) becomes false.

If the coefficient a_2 ($a_2 = -2A_1$) is positive, the value $\sigma = -2$ will serve. In fact (7) yields in this case (when both $a_2 \ge 0$ and $a_4 > 0$)

the more accurate inequality

$$a_4 \leq 4 - (2 - a_2) - \tfrac{1}{6}(2 - a_2)^2(7 + a_2).$$

The choice $\sigma = -2$ in (10) also settles (9) in the case in which $\mathrm{Re}\{A_1^3\} \leq 0$ and $\rho^2 \leq \frac{4}{5}$. If $\mathrm{Re}\{A_1^3\} \geq 0$ (for all $\rho \in [0, 1]$), (9) follows by setting $\sigma = i$ in (10) and using (8).

If $A_1^3 = -\rho^3 e^{-2i\delta}$, it thus remains to consider the case $\frac{4}{5} < \rho \leq 1$, $\sin\delta \leq 2^{-\frac{1}{2}}$. Here, a suitable choice for σ is $\sigma = -2(\cos\delta)^{-1}e^{i\delta} = -2 - 2i\tan\delta$. An elementary computation shows that, with this value of σ, (10) implies

$$0 \leq -U + \tfrac{5}{3}(1 - \rho^3)\sin^2\delta + R'\sin 2\delta - \tfrac{5}{6}[T + \rho^3]\sin^2 2\delta, \tag{11}$$

where

$$U = \tfrac{1}{2}(1 - \rho^2)[1 + \tfrac{7}{6}(1 - \rho)^2(1 + 2\rho)], \quad \phi = R + iR',$$

and

$$T = \frac{5}{4}\,\frac{(1 + \rho^3)^2}{1 + 12\rho^2}.$$

The further manipulation depends on whether the inequality $\sin^2\delta \leq \frac{18}{25}(1 - \rho^2)$ does or does not hold. In the first case, we obtain from (11) (by maximizing over $\sin 2\delta$) and (8) that

$$0 \leq -U + \tfrac{6}{5}(1 - \rho^3)(1 - \rho^2) + \frac{3\rho^2(1 - \rho^2)}{5(T + \rho^3)},$$

and this is easily shown to be false for $\frac{4}{5} \leq \rho < 1$.

In the second case, we start from the inequality $R \geq \frac{5}{6}[1 - \rho^3 + 2\rho^3\sin^2\delta]$ ($\phi = R + iR'$) which follows from (10) for $\sigma = 0$. Combining this with (8), we find that $R' \leq \frac{5}{6}\rho^3 \sin 2\delta$ if $\sin^2\delta \geq \frac{18}{25}(1 - \rho^2)$. Hence, (11) implies that

$$0 \leq -U - \tfrac{5}{3}[2T\cos^2\delta - (1 - \rho^3)]\sin^2\delta,$$

and this is false for $\sin^2\delta \leq \frac{1}{2}$, $\frac{4}{5} \leq \rho < 1$. Thus, (9) is proved in all cases.

References

1. Z. Charzynski and M. Schiffer. A new proof of the Bieberbach conjecture for the fourth coefficient, Arch. Rational Mech. Anal. 5 (1960), 187-93.

2. P. R. Garabedian and M. Schiffer. A proof of the Bieberbach conjecture for the fourth coefficient, J. Rational Mech. Anal. 4 (1955), 427-65.

3. P. R. Garabedian, G. G. Ross and M. Schiffer. On the Bieberbach conjecture for even n, J. Math. Mech. 14 (1965), 975-89.

4. K. Loewner. Untersuchungen über schlichte konforme Abbildungen des Einheitskreises, Mat. Ann. 89 (1923), 103-21.

5. M. Ozawa. On the Bieberbach conjecture for the sixth coefficient, Kodai Math. Sem. Rep. 21 (1969), 97-128.

6. R. Pederson. A Proof of the Bieberbach conjecture for the sixth coefficient, Arch. Rational Mech. Anal. 31 (1968), 331-51.

7. R. Pederson and M. Schiffer. A proof of the Bieberbach conjecture for the fifth coefficient, Arch. Rational Mech. Anal. 45 (1972), 161-93.

Carnegie-Mellon University,
Pittsburgh, U. S. A.

COMPLETENESS QUESTIONS AND RELATED DIRICHLET POLYNOMIALS

D. J. NEWMAN

Taking our cue from the function $e^{i\theta}$ (on the circle) we observe that, for certain $f(\theta)$, the collection $\{f(n\theta), -\infty < n < \infty\}$ can span certain whole function classes. Indeed, for $f(\theta) = e^{i\theta}$ it is the lesson of Fourier series that this collection is complete in L^p, $1 \leq p < \infty$ and also in C, the continuous functions, in the Sup norm. The completeness questions to which our title refers are those of determining just when $f(\theta)$ has this spanning property and in just which norms.

In this generality the problem seems altogether too ambitious and one has sufficient difficulty even under the simplifying assumption that $f(\theta)$ is a trigonometric polynomial. Within this framework it is not difficult to reduce the question further to the case of analytic polynomials in $z = e^{i\theta}$, and by so doing we obtain the following reformulation:

Let $P(z)$ be a polynomial and $1 \leq p \leq \infty$. Determine under what conditions the combinations $a_1 P(z) + a_2 P(z^2) + \ldots a_k P(z^k)$ approximate the function z arbitrarily closely in H^p.

At this point one is able to introduce a particularly attractive formalism by use of Dirichlet series. For suppose we make the association

$$\mathfrak{D} : \alpha_1 z + \alpha_2 z^2 + \ldots \alpha_n z^n \rightarrow \alpha_1 + \frac{\alpha_2}{2^s} + \ldots \frac{\alpha_n}{n^s}.$$

We then easily see that $\mathfrak{D}(P(z^\nu)) = \nu^{-s}\mathfrak{D}(P(z)) = \nu^{-s}D(s)$, call it, so that $a_1 P(z) + a_2 P(z^2) + \ldots a_k P(z^k)$ being 'near' z is equivalent to

$$(a_1 + \frac{a_2}{2^s} + \ldots \frac{a_k}{k^s})D(s)$$

being 'near' 1.

Our whole problem thus takes on a more familiar tone. We are simply asking when a given Dirichlet polynomial, $D(s)$, is 'invertible'.

The only nuisance, of course, is the fact that the norms, which are so pleasant in the z variable become rather complicated in the transformed variable, s.

To no one's surprise the case of L^2 is especially simple. After all the operator $\mathfrak{D}$ is just the Mellin, or equivalently, Fourier, transform and the pleasant norm remains pleasant! The L^2 condition is simply that the Dirichlet polynomial $D(s)$ have no zeros in the half plane $\mathrm{Re}\, s > 0$.

(This fact, together with Turan's result connecting the Riemann Hypothesis with the zeros of the partial sums of $\zeta(s)$, gives an approach to R. H. via completeness problems. Still another contribution to Pólya's list of 'How not to solve the Riemann Hypothesis.')

Exploration of the 'rock bottom' simplest $P(z)$, $z + cz^2$, yields the following:

Completeness is obtained for p, $1 \leq p < \infty$, if and only if $|c| \leq 1$.

Completeness is obtained for $p = \infty$ if and only if $|c| < 1$.

We are thereby led to conjecture, for general $P(z)$,

Completeness for p, $1 \leq p < \infty$, iff $D(s)$ has no zeros in $\mathrm{Re}\, s > 0$.

Completeness for $p = \infty$ iff $|D(s)|$ is bounded below in $\mathrm{Re}\, s > 0$.

Certain cases of this conjecture have been verified but the general result seems to hinge on some delicate facts regarding the Mellin transform. It would go a long way, for example, if one could prove that the L^p norms of $P(z)$ are bounded by the Sup norm of $D(s)$ (in $\mathrm{Re}\, s > 0$).

Finally we will end with an amusing by-product of these investigations.

Theorem. If $P(z, w)$ is a polynomial with a zero in $|z| < 1$, $|w| < 1$ then it already has a zero there with $z = w^{\sqrt{2}}$ (some determination).

Yeshiva University,
New York, U. S. A.

ON THE BOUNDARY BEHAVIOUR OF NORMAL FUNCTIONS

CH. POMMERENKE

The function $f(z)$ meromorphic in $|z| < 1$ is called normal (Lehto and Virtanen 1957) if

$$\sup_{|z|<1} (1 - |z|^2) \frac{|f'(z)|}{1 + |f(z)|^2} < \infty .$$

For instance, every simple automorphic function for a finitely generated Fuchsian group of the first kind is normal. Bagemihl and Seidel (1961) have shown that an analytic normal function has (possibly infinite) angular limits on a dense subset of $|z| = 1$. The above automorphic functions have angular limits only at the parabolic fixed points, hence only on a countable set.

Hayman (1955) has proved that every normal analytic function satisfies

$$\log^+ |f(z)| = O(\frac{1}{1 - |z|}) \qquad (|z| \to 1 - 0).$$

Theorem. Let $f(z)$ be analytic and normal in $|z| < 1$ and let

$$(*) \qquad \log^+ |f(z)| = o(\frac{1}{1 - |z|}) \qquad (|z| \to 1 - 0).$$

Then $f(z)$ has angular limits on an uncountably dense subset of $|z| = 1$.

This result of Lohwater and the author improves a theorem of R. L. Hall (1968) who made the assumption that either (*) and also $f(z) \neq 0$ $(|z| < 1)$ holds or that $|f(z)| = o((1 - |z|)^{-1})$ holds.

The idea of the proof is to consider the components G of $\{z : |f(z)| > m\}$ and then form the bounded function $g(s) = mf(\phi(s))^{-1}$ in $|s| < 1$ where $\phi(s)$ is the mapping function onto the universal covering surface of G. The assumption (*) is used to exclude the possibility of point masses in the representation of the function $g(s)$.

The assumption that $f(z)$ is normal is used, in particular, to show that each point on ∂G is the image of at most countably many points on $|s| = 1$.

Technische Universität,
Berlin, Germany

JOINT APPROXIMATION IN THE COMPLEX DOMAIN

L. A. RUBEL

This talk is a survey of recent results, none of them yet published (March 1973). A general problem of joint approximation is specialized in several different ways to yield interesting and challenging problems in hard analysis. Let E_α, $\alpha \in A$, be a collection of topological spaces (usually there are just two E_α) and let E be a space, with mappings $\rho_\alpha : E \to E_\alpha$ for each $\alpha \in A$. Put on E the coarsest topology τ that makes all the ρ_α continuous. For each $\alpha \in A$, let P_α be a dense subset of E_α, and let $P = \bigcap_{\alpha \in A} \rho_\alpha^{-1}(P_\alpha)$. When is P dense in E?

To specialize, let $D = \{z : |z| < 1\}$ be the open unit disc, and let F be a relatively closed subset of D. Let E_1 be the space of bounded analytic functions on D in the 'topology' of bounded convergence. That is, $f_n \to f$ if (a) f_n is uniformly bounded, and (b) $f_n \to f$ uniformly on compact subsets of D. Let E_2 be the space of continuous functions on F in the narrow topology. That is, $f_n \to f$ if and only if (a) $\|f_n\|_F \to \|f\|_F$ and (b) $f_n \to f$ uniformly on compact subsets of F. Here, $\| \; \|_F$ denotes the supremum on F. Let P_1 and P_2 be the ordinary algebraic polynomials, restricted to D and to F, respectively. When P is dense in E, we say that F is a Farrell set. In other words, F is a Farrell set precisely when for every bounded analytic function f in D, we can find a sequence $\{p_n\}$ of polynomials so that

(i) $\|p_n\|_D \le M < \infty$,

(ii) $\|p_n\|_F \to \|f\|_F$, and

(iii) $p_n(z) \to f(z)$ for all $z \in D$.

A. Stray has shown that F is a Farrell set if and only if almost every (in the sense of Lebesgue measure on the unit circle ∂D) point of $\overline{F} \cap \partial D$ is a non-tangential limit point of points of F.

In a similar vein, consider the space $U(F)$ of all functions f that are analytic on D and uniformly continuous on F, in the topology of uniform convergence on all sets of the form $K \cup F$, where K ranges over the compact subsets of D. When the polynomials are dense in $U(F)$, we say that F is a Mergelyan set. As a beginning, L. A. Rubel, A. L. Shields, and B. A. Taylor showed that there are two Mergelyan sets whose union is not a Mergelyan set, and two whose intersection is not. Their proof of this last assertion uses the following result about analytic functions. Let A be the space of continuous functions on the closed unit disc $\bar{D}$ that are analytic in D. Let

$$\omega(\delta, f) = \sup \{ |f(z) - f(w)| : |z - w| \leq \delta, z, w \in D \}$$

$$\tilde{\omega}(\delta, f) = \sup \{ |f(z) - f(w)| : |z - w| \leq \delta, z, w \in \partial D \}.$$

We prove that there is an absolute constant C such that $\omega(\delta, f) \leq C\tilde{\omega}(\delta, f)$, but that the best possible constant C is greater than 1.07. Later, A. Stray found necessary and sufficient geometrical conditions that a set be a Mergelyan set. He proved that F is a Mergelyan set if and only if, for every compact set K in D, $D \cap \mathcal{H}_p(K \cup F) = \mathcal{H}_{U(F)}(K \cup F)$, where $\mathcal{H}_p$ is the polynomial hull, and $\mathcal{H}_{U(F)}$ is a hull relative to $U(F)$. Space does not allow a more precise description. The polynomial hull is easily characterized geometrically and $\mathcal{H}_{U(F)}$ is characterized geometrically by Stray in terms of accessibility of boundary points of the set concerned.

Finally, there is the work of my student S. Venkateswaran on uniform approximation of continuous functions on compact sets by splines of polynomials, and on uniform approximation of continuous functions on closed sets by splines of entire functions.

Definition. Let $\{E_i\}$, $i = 1, 2, \ldots$, be a countable class of compact nowhere dense sets such that $\bigcup_{i=1}^{\infty} E_i = E$ is compact. A function g on E is called a spline of polynomials on $\{E_i\}$ if g restricted to E_i is, for each i, the restriction of a polynomial to E_i.

If we replace compact sets by closed sets and polynomials by entire functions, we arrive at the definition of a spline of entire functions.

Theorem α. Let $\mathcal{E} = \{E_i\}_{i=1}^{\infty}$ be a countable class of compact nowhere dense sets such that $\bigcup_{i=1}^{\infty} E_i = E$ is compact. Then every continuous function on E is uniformly approximable by splines of polynomials on $\{E_i\}$ if and only if $\mathbf{C}^\wedge \backslash H(\mathcal{B})$ is connected for every maximal chainable class $\mathcal{B}$ in the saturated hull $\mathcal{A}(\mathcal{E})$.

Theorem β. Let $\mathcal{E} = \{E_i\}_{i=1}^{\infty}$ be a countable class of closed nowhere dense sets in the complex plane such that $\bigcup_{i=1}^{\infty} E_i = E$ is closed. Then every continuous function on E is uniformly approximable by splines of entire functions on $\{E_i\}$ if and only if $\mathcal{H}(\mathcal{B})$ is an Arakelian set for every maximal chainable class $\mathcal{B}$ in the saturated hull $\mathcal{A}(\mathcal{E})$.

Some explanation of terminology is needed. Here $\mathbf{C}^\wedge$ is the Riemann sphere.

Definition. If $\mathcal{A}$ is a class of closed sets and if $\mathcal{B}$ is a subclass of $\mathcal{A}$, we call $\mathcal{B}$ chainable in $\mathcal{A}$ if for any two sets A and B in $\mathcal{B}$, we can find sets $A_0 = A, A_1, A_2, \ldots, A_{n-1}, A_n = B$ in $\mathcal{A}$ such that $A_i \cap A_{i+1}$ has a finite limit point for $i = 0, 1, 2, \ldots, n-1$.

It is now clear what is meant by 'maximal chainable'. If $\mathcal{B}$ is a maximal chainable class in $\mathcal{A}$, then we denote by $H(\mathcal{B})$ the closure of the union of all sets in $\mathcal{B}$.

Now let $\mathcal{E}$ be a class of closed sets, and let $\mathcal{A}^0 = \mathcal{E}$. Further, let

$$\mathcal{A}^\eta = \{H(\mathcal{B}) : \mathcal{B} \text{ is a maximal chainable class in } \mathcal{A}^\xi \text{ for some } \xi < \eta\}.$$

Here, ξ and η are ordinal numbers. It is shown that there is a first ordinal η_0 such that $\mathcal{A}^{\eta_0} = \mathcal{A}^{\eta_0+1}$

Definition. The saturated hull $\mathcal{A}(\mathcal{E})$ is defined as the class of all sets in $\mathcal{A}^\eta$ for all $\eta \leq \eta_0$. Finally, a set E in the complex plane is called an Arakelian set if it is nowhere dense, if $\mathbf{C}^\wedge \backslash E$ is connected, and if $\mathbf{C}^\wedge \backslash E$ is locally arcwise connected.

The general problem gives rise to many other special problems of analytical interest. The hope I once had of finding general principles that solve all of them is very dim now in the face of the great diversity of the now known answers to some of the interesting special cases.

University of Illinois,
Urbana, U. S. A.

SOME LINEAR OPERATORS IN FUNCTION THEORY

T. B. SHEIL-SMALL

1. Consider the general class of continuous linear operators

$$\Lambda : \mathcal{A} \to \mathcal{A}$$

where $\mathcal{A}$ denotes the space of functions analytic in $|z| < 1$ with the topology of local uniform convergence. If we write

$$\Lambda z^n = \tau_n(z) \qquad (n = 0, 1, 2, \ldots)$$

then for $f(z) = \sum_0^\infty a_n z^n \in \mathcal{A}$ we have

$$(\Lambda f)(z) = \sum_0^\infty a_n \tau_n(z)$$

where

$$\limsup_{n\to\infty} (M(r, \tau_n))^{1/n} < 1 \qquad (0 < r < 1)$$

and conversely every sequence $\{\tau_n(z)\}$ in $\mathcal{A}$ satisfying this last condition generates a continuous linear operator Λ. Thus the class of operators Λ is associated with the class of functions $H(z, \zeta)$ analytic in $|z| < 1$, $|\zeta| \le 1$ expanded in the form

$$H(z, \zeta) = \sum_0^\infty \tau_n(z)\zeta^n.$$

We say Λ is a <u>convexity preserving operator</u> if for each $f \in \mathcal{A}$ the range of values of Λf lies in the closed convex hull of the range of f. It is easily seen that Λ is convexity preserving if, and only if, $\tau_0(z) = 1$ and

$$\operatorname{Re}\left(\sum_0^\infty \tau_n(z)\zeta^n\right) > \tfrac{1}{2} \quad (|z| < 1, \ |\zeta| \le 1).$$

2. Consider now the class of operators satisfying the condition

$$\Lambda z^n = z^n \sigma_n(z) \qquad (n = 1, 2, \ldots)$$

each $\sigma_n \in \mathcal{A}$. If we set

$$K(z, \zeta) = \sum_0^\infty \sigma_n(\zeta) z^n$$

then $K(z, \zeta)$ is analytic in the polydisc $|z| < 1$, $|\zeta| < 1$ and conversely every such function generates an operator of this type. If $f(z) = \sum_0^\infty a_n z^n \in \mathcal{A}$ we have

$$\Lambda f(z) = K(z, \zeta) \circledast f(z) = \sum_0^\infty a_n \sigma_n(z) z^n .$$

Our operator is thus a generalised convolution or Hadamard product operator, the simple convolution operator occurring in the case when K is a function of z alone. Another example of a g. c. operator is the subordination operator obtained as follows: let $w(z) \in \mathcal{A}$ and satisfy $w(0) = 0$, $|w(z)| < 1$. Then for $f \in \mathcal{A}$

$$f(w(z)) = (K \circledast f)(z)$$

where

$$K(z, \xi) = \sum_0^\infty \left(\frac{w(\xi)}{\xi}\right)^n z^n$$

$$= \frac{1}{1 - \frac{w(\xi)}{\xi} z}$$

This is trivially a convexity preserving operator, and in general the necessary and sufficient condition that K be a convexity preserving g. c. operator is that $K(0, \xi) = 1$ and

$$\operatorname{Re} K(z, \xi) > \tfrac{1}{2} \qquad (|z| < 1, \ |\xi| < 1).$$

We also consider bound preserving g. c. operators. This is defined by the condition

$$* \qquad M(r, K \circledast f) \leq M(r, f) \qquad (0 < r < 1)$$

and every $f \in \mathcal{A}$. Clearly

convexity preservation $\Rightarrow$ bound preservation

When the condition * holds for every $f \in \mathcal{B} \subset \mathcal{A}$ we say that K is bound preserving over $\mathcal{B}$.

If $g \in \mathcal{A}$ satisfies $|g(z)| \leq 1$ in $|z| < 1$ then

$$\Lambda f = gf \qquad (f \in \mathcal{A})$$

defines a bound preserving g. c. operator over $\mathcal{A}$. Multiplication by a fixed function g may be called a simple multiplication operator.

3. The following result generalises the subordination invariance theorem proved in [2].

Theorem. Let J be a convexity preserving g. c. operator and suppose that K is a g. c. operator bound preserving over one of the following four classes:

(i) $\mathcal{A}$

(ii) $\mathcal{A}_0$ = subset of $\mathcal{A}$ vanishing at 0

(iii) $\mathcal{A}^{(n)}$ = set of polynomials of degree $\leq n$

(iv) $\mathcal{A}_0^{(n)}$ = subset of $\mathcal{A}^{(n)}$ vanishing at 0.

Define $H(z, \xi) = J \circledast K_\xi$ where $K_\xi = K(z, \xi)$ with ξ regarded as fixed. Then H is a g. c. operator bound preserving over the same class as K.

The operator here described sending $K(z, \xi) \rightarrow H(z, \xi)$ is a linear operator on $\mathcal{A}^2$, the space of functions analytic in the polydisc. I intend to study more closely those operators which preserve the structure of K.

4. Integral mean estimates

Theorem. Let K be a bound preserving g. c. operator over $\mathcal{A}$ and let $f \in \mathcal{A}$. Then if $0 < r < R < 1$, $\exists\, A_K(r, R)$ independent of f and satisfying

$$A_K(r, R) \leq \frac{R + r}{R - r}$$

such that for each $p \geq 1$

$$\int_0^{2\pi} |(K \circledast f)(re^{i\theta})|^p d\theta \leq A_K(r, R) \int_0^{2\pi} |f(Re^{i\theta})|^p d\theta .$$

Furthermore we can take $A_K(r, R) = 1$ *(and hence* $R = r$*) in the following cases:*

(i) K *is convexity preserving*

(ii) K *is a simple convolution operator*

(iii) K *is a simple multiplication operator.*

Corollary. *Let* K *be of one of the types (i)-(iii) and let* $f(z) = \sum_0^\infty a_n z^n$, $K \circledast f = \sum_0^\infty b_n z^n$ *where* $f \in \mathfrak{A}$. *Then for each* $n \geq 0$

$$\sum_{k=0}^{n} |b_k|^2 \leq \sum_{k=0}^{n} |a_k|^2 .$$

Proof. Writing $f_n = e_n * f$, $e_n = 1 + z + \ldots + z^n$ we have

$$e_n * (K \circledast f) = e_n * (K \circledast f_n) .$$

Hence

$$\sum_0^n |b_k|^2 r^{2k} = \frac{1}{2\pi} \int_0^{2\pi} |e_n * (K \circledast f)(re^{i\theta})|^2 d\theta$$

$$= \frac{1}{2\pi} \int_0^{2\pi} |e_n * (K \circledast f_n)(re^{i\theta})|^2 d\theta$$

$$\leq \frac{1}{2\pi} \int_0^{2\pi} |(K \circledast f_n)(re^{i\theta})|^2 d\theta$$

$$\leq \frac{1}{2\pi} \int_0^{2\pi} |f_n(re^{i\theta})|^2 d\theta = \sum_0^n |a_k|^2 r^{2k}.$$

The theorem generalises for $p \geq 1$ Littlewood's well-known subordination inequality. The corollary generalises Rogosinski's well-known deduction from Littlewood's result.

5. Our theorems all apply to some recently obtained non-trivial convexity preserving g. c. operators arising out of the proof of the Pólya-Schoenberg conjecture. Let (ϕ, g) be a pair of functions in $\mathfrak{A}_0$ such that either ϕ is convex and g starlike or both ϕ and g are starlike of order $\frac{1}{2}$. It was shown in [1] that the operator

$$\Lambda F = \frac{\phi * gF}{\phi * g} \qquad (F \in \mathcal{A})$$

is convexity preserving. It is clearly a g. c. operator. It is easily shown that a necessary and sufficient condition for a couple (ϕ, g) in $\mathcal{A}_0 \times \mathcal{A}_0$ to define a convexity preserving g. c. operator as described is that

$$\phi(z) * \frac{1 + \alpha z}{1 + \beta z} g(z) \neq 0 \qquad (0 < |z| < 1)$$

for every α and β satisfying $|\alpha| = |\beta| = 1$.

References

1. St. Ruscheweyh and T. Sheil-Small. Hadamard products of schlicht functions and the Pólya-Schoenberg conjecture, Comm. Math. Helv., 48 (1973).
2. T. Sheil-Small. On the convolution of analytic function, J. Reine angew. Math., 258 (1973), 137-52.

University of York,
York, England.

ANALOGUES OF THE ELLIPTIC MODULAR FUNCTION IN R^3

URI SREBRÖ

1. Introduction

The purpose of this talk is to present mappings g of the upper half space $H = \{x \in R^3 : x_3 > 0\}$ into R^3 which resemble in some respects the elliptic modular function and raise some questions in the theory of quasiregular mappings. Each of the mappings g has the following properties:

(1) g is continuous, discrete, open, sense-preserving and has a bounded dilatation in H.

(2) g defines a closed map of H onto $R^3 \setminus P$ for some set $P \subset R^3$ of finite cardinality.

(3) there exists a discrete group G of Möbius transformations acting on H with a non-compact fundamental domain of finite hyperbolic volume in H such that $g \circ A = g$ for all $A \in G$.

(4) g has no limit at any point $b \in \partial H$.

(1) says that g is quasiregular in the sense of Martio, Rickman and Väisälä [1]. This means that $g \in ACL^3$ and $|g'(x)|^3 < KJ(x, g)$ a. e. in H for some $K \in [1, \infty)$, where $|g'(x)|$ denotes the sup norm of the formal derivative $g'(x)$, and $J(x, g) = \det g'(x)$.

The class of quasiregular mappings in R^3 is a reasonable generalization of holomorphic functions in **C**; and with (2)-(4) we may consider the mappings g as analogues of the elliptic modular function. However, contrary to the elliptic modular function, none of the mappings g that are constructed here is a local homeomorphism. Martio and I show in [4] that no quasiregular mapping in R^n, $n \geq 3$, which satisfies (3) is a local homeomorphism. It thus follows that no mapping in R^n, $n \geq 3$, can be completely analogous to the elliptic modular function.

The mappings g which are presented here are examples of automorphic mappings in R^3. A comprehensive study of automorphic mappings in $\overline{R}^n = R^n \cup \{\infty\}$, $n \geq 2$, will appear in [4].

2. The construction of g

Map the region $\overline{D} = \{x \in \overline{H} : x_1 \geq \frac{1}{2},\ x_1 - 1 \leq \sqrt{3}x_2 \leq 1 - x_1,\ |x| > 1\}$ (its projection on ∂H is described by the shaded region in Figure 4 below) homeomorphically onto $\overline{H}$, so that the points $a_1 = (1, 0, 0)$, $a_2 = (-\frac{1}{2}, \frac{\sqrt{3}}{2}, 0)$, $a_3 = (-\frac{1}{2}, \frac{-\sqrt{3}}{2}, 0)$ and $a_4 = \infty$ are kept fixed, the edges of D correspond to line segments in ∂H and the restriction of the map to $D = \operatorname{int} \overline{D}$ is quasiconformal. Now, extend the map by means of reflections in the faces of D and ∂H respectively. The angle between any two adjacent faces of D is $\frac{\pi}{3}$, thus, cf. [5], the group Γ generated by the reflections $T_1, \ldots, T_4$ in the four faces of D is discrete and D is a fundamental domain for Γ. Therefore we can continue and extend the mapping, by means of all possible reflections, to end with a well defined mapping $g : H \to R^3 \setminus \{a_1, a_2, a_3\}$. It is not hard to verify that g satisfies (1)-(3) where G is the subgroup of all sense preserving elements of Γ. Note that $\operatorname{int}(T_1(D) \cup \overline{D})$ is a fundamental domain for G satisfying (3). Finally every point $b \in \partial H$ is a limit point for G and $g \circ A = g$ for all $A \in G$; consequently (4) holds.

Clearly, g has a non-void branch set B_g. Actually, $x \in B_g$ iff x is G-equivalent to a point in an edge of D; and gB_g is the union of six line segments joining the excluded points $a_1, \ldots, a_4$.

3. Further examples

Figures j, j=1, ..., 5, below, define hyperbolic polyhedra D_j in H. The circle and the lines, in each figure, describe the trace of the faces of D_j in ∂H. The shaded regions describe the projection of D_j on ∂H, $j = 1, \ldots, 5$. The angle between any two adjacent faces of D_j, $j = 1, \ldots, 5$, is a submultiple of π; hence we may apply the method of section 2 above, starting now with D_j instead of D to obtain mappings $g_j : H \to R^3$, $j = 1, \ldots, 5$, which satisfy (1)-(4) with card $(\overline{R}^3 \setminus g_j(H)) = j$. Here $\overline{R}^3 = R^3 \cup \{\infty\}$.

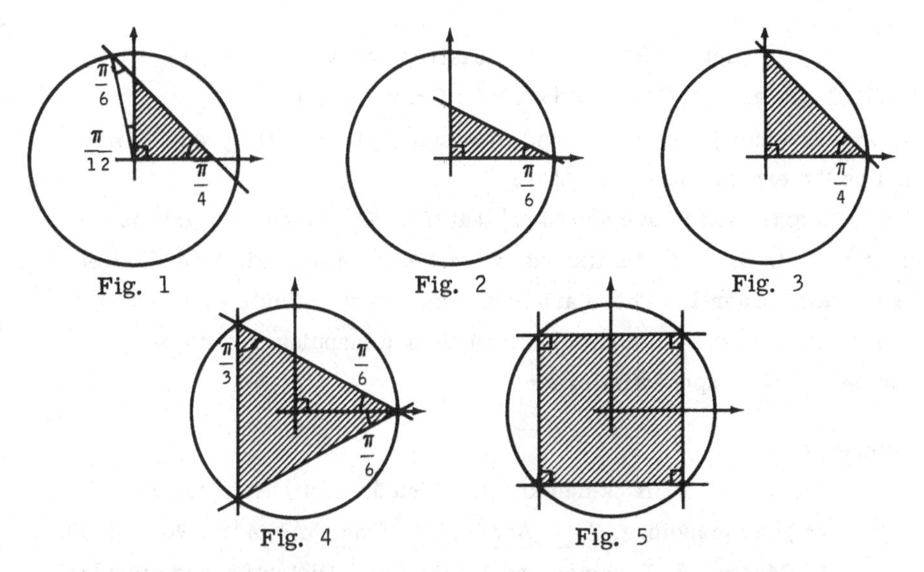

Fig. 1 Fig. 2 Fig. 3

Fig. 4 Fig. 5

The mapping g_4 which corresponds to Fig. 4 is described above in section 2, while the mapping g_3 (corresponding to Fig. 3) is studied in detail in [4, 4. 3].

4. An existence problem

Let G be a discrete Möbius group acting on H. Does G possess automorphic mappings $g : H \to \overline{R}^3$ of bounded dilatation?

So far we have only a partial answer. Martio and I have proved [4, 4. 4] that if H/G is of finite hyperbolic volume, the answer is yes. We were not able to prove the existence or non-existence of automorphic mappings for groups G with quotation space H/G of infinite volume.

5. Further problems in the theory of quasiregular mappings

(1) Does the Picard theorem hold for quasiregular mappings in R^3? If not, then how big is the exceptional set $R^3 \setminus f(R^3)$?;

Martio, Rickman and Väisälä [2] have proved that $R^3 \setminus f(R^3)$ is of zero capacity and Zoric [6] has constructed a quasiregular mapping of R^3 onto $R^3 \setminus \{0\}$.

(2) Let $f : R^3 \to R^3$ be quasiregular with $f(x + c_1) = f(x)$ for all $x \in R^3$, $D = \{x \in R^3 : 0 \le x_i < 1\}$, $N(y) =$ cardinality $D \cap f^{-1}(y)$ and $N = \sup N(y)$ over all $y \in R^3$. Can N be finite?

It can be shown that if $N < \infty$, then $\lim f(x) = \infty$ as $x \to \infty$ in D and that $f(R^3) = R^3$. Our guess [3] is that always $N = \infty$.

(3) Let $f : R^3 \to R^3$ be quasiregular with $f(x + e_i) = f(x)$ $i = 1, 2$ for all $x \in R^3$, $D = \{x \in R^3 : 0 \leq x_i < 1, i = 1, 2\}$, $N(y) =$ cardinality $D \cap f^{-1}(y)$ and $N = \sup N(y)$ over all $y \in R^3$. What can be the exponential order of f?

Martio and I have shown [3] that if $N < \infty$ then $\lim f(x)$ as $x_3 \to \infty$, x in D and $\lim f(x)$ as $x_3 \to -\infty$, x in D, exist and f is of exponential order 1. There are mappings f with infinite exponential order and $N = \infty$. We do not know whether a mapping f with $N = \infty$ can be of finite exponential order.

References

1. O. Martio, S. Rickman and J. Väisälä. Definitions for quasiregular mappings, Ann. Acad. Sci. Fenn., A1, 448 (1969), 1-40.
2. O. Martio, S. Rickman and J. Väisälä. Distortion and singularities of quasiregular mappings, Ann. Acad. Sci. Fenn., A1, 465 (1970), 1-13.
3. O. Martio and U. Srebrö. Periodic quasimeromorphic mappings in R^n, J. d'Analyse Mat. (to appear).
4. O. Martio and U. Srebrö. Automorphic quasimeromorphic mappings in R^n, (to appear).
5. E. R. Vinberg. Discrete linear groups generated by reflections, Math. USSR Izv., vol. 5 (1971), 1083-119.
6. V. A. Zoric. The theorem of M. A. Lavrentiev on quasiconformal mappings in space, Mat. Sbornik, 74 (1967), 417-44 (Russian).

Technion,
Haifa, Israel

ON SOME PHENOMENA AND PROBLEMS OF THE POWERSUM-METHOD

PAUL TURAN

Let $g(\nu) = \sum_{j=1}^{n} b_j z_j^{\nu}$ a generalised powersum of the complex numbers $z_1, \ldots, z_n$ and

$$\xi = (z_1, \ldots, z_n) \in A \text{ means } \min_j |z| = 1$$

$$\xi = (z_1, \ldots, z_n) \in B \text{ means } |z_j| = 1, \ j = 1, 2, \ldots, n.$$

Let further for $k \geqq n$

$$M_1(k) = \inf_{b_j} \inf_{\xi \in A} \max_{\nu=1, \ldots, k} \left|\frac{g(\nu)}{g(0)}\right| .$$

The sharp form of the so-called first main theorem gives

$$(1) \qquad M_1(n) = \frac{1}{2^n - 1} .$$

In contrast to various investigations, where mainly

$$\inf_{b_j} \inf_{\xi \in A} \max_{\nu=m+1, \ldots, m+n} \left|\frac{g(\nu)}{g(0)}\right|$$

for variable m was treated and applied, here $M_1(k)$ for variable k is investigated as well as

$$M_2(k) = \inf_{b_j > 0} \inf_{\xi \in A} \max_{\nu=1, \ldots, k} \left|\frac{g(\nu)}{g(0)}\right|$$

and

$$M_3(k) = \inf_{b_j > 0} \inf_{\xi \in B} \max_{\nu=1, \ldots, k} \left|\frac{g(\nu)}{g(0)}\right| .$$

The investigation of $M_1(k)$ leads to the extremal problem to determine

$$\sup_{\xi \in A} \min_{x} \sum_{\nu=1}^{k} |d_\nu| = U(k, n),$$

where the quantities $d_\nu = d_\nu(x, \xi)$ are defined by

$$\prod_{j=1}^{n} \left(1 - \frac{z}{z_j}\right)(1 + x_1 z + \dots + x_{k-n} z^{k-n}) = 1 + \sum_{\nu=1}^{k} d_\nu z^\nu ,$$

and the inequality

$$M_1(k) \geqq \frac{1}{U(k, n)}$$

is pointed out. More specially and explicitly the case $k = nl$ (l integral) is treated through the extremal problem of G. Halász which asks for the value of

$$H_l = \min_{\Pi_l} \max_{|z|=1} |\pi_l(z)| ,$$

where Π_l consists of the polynomials $\pi_l(z)$ of lth degree satisfying the restrictions

$$\pi_l(0) = 1, \qquad \pi_l(1) = 0.$$

A method of Rahman-Stenger is sketched for the estimation

$$H_l \leqq e^{2/l}$$

which leads quickly to

$$M_1(l n) > \frac{c_1}{\sqrt{l n}} e^{-4n/l}$$

and especially for $l = n$

$$\text{(2)} \qquad M_1(n^2) \geqq \frac{c_2}{n},$$

where $c_1, \dots$ are positive numerical constants (and later $c(\varepsilon)$ depends only upon ε). This gives a much bigger lower bound than (1) and its order is probably best possible.

Turning to $M_2(k)$ we have of course

$$(3)\qquad M_2(n) \geqq \frac{1}{2^n - 1}, \qquad M_2(n^2) \geqq \frac{c_2}{n}$$

and also, using Dirichlet's theorem,

$$M_2(5^n) \geqq \cos\frac{2\pi}{5}$$

independently of n. All three inequalities hold obviously replacing M_2 by M_3. In order to motivate the interest of a closer investigation of $M_3(k)$ a brief sketch of the proof was given that the inequality

$$(4)\qquad M_3(n^B) \geqq \frac{\log^{3/2} n}{\sqrt{n}}$$

with a fixed positive B and $n > n_0(B)$ implies that Riemann's conjecture is incorrect. After some remarks concerning the cases, when the Riemann conjecture is known to be true, indications were given for the modification of the proof to support the contention that replacing (4) by the inequality

$$(5)\qquad M_3(n^B) \geqq c(\varepsilon)\frac{\log^{\varepsilon} n}{\sqrt{n}}$$

for $n > n_0(B, \varepsilon)$ ($\varepsilon > 0$, arbitrarily small) the same conclusion can be drawn. As to $M_3(k)$ the recent inequalities

$$(6)\qquad M_3((1+\varepsilon)n) \geqq \frac{c(\varepsilon)}{\sqrt{n}}$$

$$(7)\qquad M_3(n^B) \leqq 9B\sqrt{\frac{\log n}{n}}, \qquad B \geqq 2$$

due respectively to D. F. Newman and R. Tijdeman were mentioned.

Eötvös Lorand University,
Budapest, Hungary

MEROMORPHIC FUNCTIONS WITH LARGE SUMS OF DEFICIENCIES

ALLEN WEITSMAN

This abstract contains an outline of a method of construction due to D. Drasin and myself which yields meromorphic functions of finite order λ $(1 < \lambda < \infty;\ \lambda \neq \frac{n}{2},\ n = 2, 3, 4, \ldots)$ having large sums of deficiencies.

The state of knowledge concerning those values of λ precluded from study is quite advanced. In fact, for order $\lambda < 1$, combined results of Valiron [5], Edrei [2], and Baernstein and Edrei [1], [3] yield the bounds

$$\sum_{a \in \hat{C}} \delta(a, f) \leq \begin{cases} 1 & (\lambda \leq \frac{1}{2}) \\ 2 - \sin \pi\lambda & (\frac{1}{2} < \lambda < 1) \end{cases} \tag{1}$$

which are seen to be sharp by well known functions. Furthermore, some 43 years ago, F. Nevanlinna [4] provided examples of orders $\lambda = 1, \frac{3}{2}, 2, \frac{5}{2}, \ldots$ for which the total deficiency sum equals 2.

The classes of functions we construct fall naturally into two families; one for which

$$\Lambda_1(\lambda) = 2 - \frac{2 \sin\{\frac{\pi}{2}(2\lambda - [2\lambda])\}}{[2\lambda] + 2 \sin\{\frac{\pi}{2}(2\lambda - [2\lambda])\}}$$

is the total sum, and one for which

$$\Lambda_2(\lambda) = 2 - \frac{2 \cos\{\frac{\pi}{2}(2\lambda - [2\lambda])\}}{[2\lambda] + 1}$$

is the total sum. Indeed, based on our investigation it seems to us likely that for functions of order $\lambda \geq 1$, Nevanlinna's classical defect relation could be refined to

$$\sum_{a \in \hat{C}} \delta(a, f) \leq \max\{\Lambda_1(\lambda), \Lambda_2(\lambda)\}.$$

The remaining values of λ would then be covered, of course, by the established bound (1).

We base our construction on a welding of portions of parabolic surfaces having finitely many branch points, with bordered surfaces which might be most appropriately called Lindelöfian ends. We require the following modified form of a theorem of Teichmüller.

Theorem A. Let $z(\zeta)$ be a homeomorphism of the finite ζ-plane to $\{|z| < R \leq \infty\}$ which for ζ outside $\{|\zeta| < \rho_0\}$ is quasiconformal. If the dilatation $p(\zeta, z)$ of $z(\zeta)$ satisfies

$$\iint_{\rho_0 < \rho < |\zeta| < 2\rho} (p(\zeta, z) - 1) \frac{d\xi d\eta}{|\zeta|^2} \to 0 \quad (\rho \to \infty),$$

where $\zeta = \xi + i\eta$, then $R = \infty$ and for any fixed $\sigma > 1$, if $\zeta_1 = \zeta_1(\rho)$, $\zeta_2 = \zeta_2(\rho)$ are any points satisfying $|\zeta_1| = \rho$, $|\zeta_2| = \sigma\rho$ we have

$$\lim_{\rho \to \infty} \frac{|z(\zeta_2)|}{|z(\zeta_1)|} = \sigma. \tag{2}$$

Very briefly our procedure is as follows:

We begin with a parabolic Riemann surface $\mathfrak{F}$ having finitely many branch points. A neighbourhood $\mathfrak{U}$ of one of the logarithmic branch points is excised and replaced by a bordered surface $\mathfrak{F}^*$ which comes from a canonical product f_μ having negative zeros, and counting function

$$n(r, 0, f_\mu) = [r]^\mu$$

for an appropriate non-integer μ; the functions f_μ are the 'Lindelöf functions'. For functions having total deficiency sum Λ_2 for instance, the Lindelöfian end $\mathfrak{F}^*$ will actually be taken as the portion of the Riemann surface for f_μ^{-1} lying over the unit disk, and μ suitably chosen in the interval $(\frac{1}{2}, 1)$.

We construct a quasiconformal mapping ψ^* of a sector in the ζ-plane to $\mathfrak{F}^*$ and weld the border of $\mathfrak{F}^*$ to $\mathfrak{F} \setminus \mathfrak{U}$. By putting together the uniformising function of $\mathfrak{F}$ restricted to $\mathfrak{F} \setminus \mathfrak{U}$ and the function ψ^* we obtain a mapping ϕ from the ζ-plane to $\mathcal{G} = (\mathfrak{F} \setminus \mathfrak{U}) \cup \mathfrak{F}^*$ which is quasiconformal outside a compact set. Letting f^{-1} be the uniformising function from $\mathcal{G}$ to the finite z-plane, the composition $z = f^{-1}(\phi(\zeta))$

becomes a homeomorphism of the ζ-plane to the z-plane satisfying the conditions of Theorem A. From (2) and the asymptotic behaviour of ϕ we then deduce the desired properties of f.

References

1. A. Baernstein II. Proof of Edrei's spread conjecture, Proc. London Math. Soc., (3) 26 (1973), 418-34.
2. A. Edrei. Sums of deficiencies of meromorphic functions, Jour. d'Analyse, 14 (1965), 79-107.
3. A. Edrei. Solution of the deficiency problem for functions of small lower order, Proc. London Math. Soc., (3) 26 (1973), 435-45.
4. F. Nevanlinna. Über eine Klasse meromorpher Funktionen, Septième Congrès Math. Scand. Oslo, 1930, pp. 81-3.
5. G. Valiron. Sur les valeurs déficientes des fonctions meromorphes d'ordre nul, C. R. Acad. Sci. Paris, 230 (1950), 40-2.

Purdue University,
Lafayette, U. S. A.

becomes a homeomorphism [illegible] plane to the [illegible] [illegible] of Theorem [illegible]

[illegible]

References

1. A. Baernstein II [illegible] [illegible] Math. [illegible] 26 [illegible]

2. [illegible]

3. A. [illegible] of [illegible]

4. [illegible]

5. [illegible]

[illegible]

ON D. J. PATIL'S REMARKABLE GENERALISATION OF CAUCHY'S FORMULA

L. C. YOUNG

According to a classical theorem of F. and M. Riesz, the values in the open unit disc of an analytic function f (z) of the Hardy class H^p are uniquely determined by the boundary values on a subset E of positive linear measure of the circumference. Patil's formula gives these values explicitly, and Cauchy's formula is the special case where the subset E reduces to the circumference itself. More recently Patil has extended his formula to functions of several complex variables on a polydisc. Patil has obtained these results by functional analysis, using operators and Toeplitz matrices. In the case of one complex variable, an elementary proof is possible, in fact it is at once suggested by the classical devices of Phragmen-Lindelöf, at least in the case where the subset in question consists of a finite sum of arcs. A simple way of passing from this special case to the general case has been given by Steve Wainger. The applications of these methods, and of the related methods of Albert Baernstein, are far-reaching.

One naturally thinks of the Goldbach problem: the Hardy-Littlewood attack amounts to estimating at the origin the nth derivative of a polynomial of degree 2n, which is well-behaved on most of the unit circumference (on the 'major arcs'). However I shall limit myself to the case in which E reduces to a single arc. The formula in this case was almost obtained by Paley-Wiener. It provides, among other things, an automatic method of analytic continuation, of the same power as the Lindelöf method of summation of a divergent series. Previously the only automatic analytic continuation was provided by the Schwarz reflexion principle, which depends on very special assumptions. Patil's analytic continuation, and the ideas behind it, can be applied to the Riemann zeta function. Both the Lindelöf hypothesis and the Riemann hypothesis can then be expressed rather simply in terms of the values of the zeta function and its logarithmic derivative along segments of the positive real axis. They amount to precise estimates of orders of magnitude of certain Fourier transforms, for large values of the arguments. All this can be done by entirely elementary methods. Previously a possible way of establishing the Riemann hypothesis, without going into the critical strip, was given by Turan: however, it was not an equivalence, and the stronger hypothesis it suggests is probably false.

References

1. A. Baernstein. Some extremal problems for univalent functions, harmonic measures and subharmonic functions, C, 6-11.
2. D. J. Patil. Representation of H^p functions, Bull. Amer. Math. Soc., 78 (1972), 617-20.
3. P. Turan. On some approximate Dirichlet polynomials in the theory of the zeta-function of Riemann, Det Kgl. Danske Vidensk. Selsk. Math.-Fys. Meddel XXIV, 17 (1948).

University of Wisconsin,
Madison, U. S. A.

ANALYTIC FUNCTIONS AND HARMONIC ANALYSIS

LAWRENCE ZALCMAN

Let μ be a (finite complex Borel) measure supported in the unit ball of $\mathbf{R}^n$. A function $u \in C(\mathfrak{D})$, where $\mathfrak{D}$ is a domain in $\mathbf{R}^n$, satisfies the generalised mean-value property with respect to μ if

$$(1) \qquad \int u(x + rt)d\mu(t) = 0 \qquad x \in \mathfrak{D}, \ 0 < r < \operatorname{dist}(x, \partial\mathfrak{D}).$$

This definition is motivated by the following two examples.

Example 1. Let $\mu = \Omega - \delta_0$, where Ω is the uniform distribution of total mass 1 on the unit sphere and δ_0 is the point mass at the origin. Condition (1) is then the classical mean-value condition, which is necessary (Gauss) and sufficient (Koebe) for u to be harmonic in $\mathfrak{D}$.

Example 2. Let $n = 2$, $d\mu = d\zeta$ on the unit circle $|\zeta| = 1$. Then (1) is both necessary (Cauchy) and sufficient (Morera, Carleman) that u be holomorphic in $\mathfrak{D}$.

In general, we ask for necessary and sufficient conditions that (1) hold. These are provided by

Theorem 1. [8] *Let* $F(z) = \int_{\mathbf{R}^n} e^{-i(z\cdot t)}d\mu(t)$ *be the Fourier-Laplace transform of the measure* μ. *Let* $F(z)$ *have the expansion* $\sum_{n=0}^{\infty} Q_n(z)$ *in terms of homogeneous polynomials. Then* $u \in C(\mathfrak{D})$ *satisfies (1) if and only if* u *is a weak solution to the system of linear partial differential equations*

$$(2) \qquad Q_n(D)u = 0 \qquad n = 0, 1, 2, \ldots .$$

Here, as usual, $z = (z_1, \ldots, z_n) \in \mathbf{C}^n$ *and* D *denotes the symbolic vector* $(-i\partial/\partial x_1, \ldots, -i\partial/\partial x_n)$.

Ingredients of the proof include a generalization of Pizzetti's formula [1, p. 288] to arbitrary measures, the Hilbert Basis Theorem, and Ehrenpreis' Fundamental Principle [3].

In certain cases, the system (2) is completely redundant and can be replaced by the single equation $P(D)u = 0$ where P is a homogeneous polynomial. This is the case, for instance, in Example 1, where $P(\xi_1, \ldots, \xi_n) = \xi_1^2 + \ldots + \xi_n^2$, and Example 2, where $P(\xi_1, \xi_2) = (\xi_1 + i\xi_2)/2$. More generally, we have

Theorem 2. [8] Let $P(\xi_1, \ldots, \xi_n)$ be a homogeneous polynomial. There exists a finite complex Borel measure μ supported in the unit ball of $\mathbf{R}^n$ such that $u \in C(\mathfrak{D})$ satisfies (1) if and only if $P(D)u = 0$ weakly. In fact, any measure of the form $\mu = P(D)T$, where T is a distribution of compact support such that $\langle T, 1\rangle \neq 0$, has this property.

Theorem 2 provides a complete answer to the question of which differential equations are equivalent to a condition of type (1); its proof uses the Nullstellensatz and the (distributional) Paley-Wiener Theorem.

In earlier work [7], we showed how the theory of mean periodic functions [5] could be used to characterize the solutions on $\mathbf{R}^2$ of the equation $(\partial/\partial z)^n(\partial/\partial\bar{z})^m f = 0$. A typical result is the following generalization of Morera's Theorem.

Theorem 3. Let $f \in L^1_{loc}(\mathbf{R}^2)$ and suppose that $\int_\Gamma f(z)dz = 0$ for almost every circle Γ having radius r_1 or r_2. Then f is (equal almost everywhere to) an entire function, so long as r_1/r_2 is not a quotient of zeroes of the Bessel function $J_1(z)$. In case r_1/r_2 is such a quotient, f need not be holomorphic anywhere and can even vanish on an open set without vanishing identically.

Polyanalytic and polyharmonic functions have similar characterizations.

Motivated by these results, let us say that the measure μ determines the differential equation $P(D)u = 0$ if there exists an at most countable (possibly empty) set of real numbers $S = S(\mu)$ such that the equation

$$\int u(x + rt)d\mu(t) = 0, \quad r = r_1, r_2$$

for almost all $x \in \mathbf{R}^n$ is equivalent to $P(D)u = 0$ (weakly) whenever $u \in L^1_{loc}(\mathbf{R}^n)$ and $r_1/r_2 \notin S(\mu)$. Then we have

Theorem 4. [8] _Let_ $P(\xi_1, \ldots, \xi_n)$ _be a homogeneous polynomial. There exists a measure_ μ _absolutely continuous with respect to volume such that_ μ _determines the differential equation_ $P(D)u = 0$.

A result with a rather different flavor is provided by

Theorem 5. [8] _Let_ $r > 0$ _be fixed and_ $0 \leq n < m$. _Suppose_ $f \in L^1_{loc}(\mathbf{R}^2)$ _and that_

$$\int f(z + re^{i\theta})e^{in\theta}d\theta = 0,$$

$$\int f(z + re^{i\theta})e^{im\theta}d\theta = 0$$

for almost all $z \in \mathbf{C}$. _Then (after modification on a set of measure_ 0) $\partial^n f/\partial \bar{z}^n = 0$.

There are corresponding results for the cases $m < n \leq 0$ and $m < 0 < n$, $m \neq -n$. The proofs depend on the important fact, first proved by C. L. Siegel [6], that the nonzero solutions of $J_\alpha(z) = 0$ are transcendental whenever α is algebraic.

Finally, we wish to mention that results similar to Theorems 3 and 4 are valid for functions defined on the hyperbolic plane Δ (the open unit disc with the Poincaré metric). Write $d\sigma(z) = 2|dz|/(1 - |z|^2)$ for invariant arc length on Δ. Then $\Gamma_r(z)$, the noneuclidean circle of radius r and center z, has ($d\sigma$) length $c_r = 2\pi \sinh r$.

Theorem 6. _Suppose_ $u \in C(\Delta)$ _and_

$$\frac{1}{c_r}\int_{\Gamma_r(z)} u(\zeta)d\sigma(\zeta) = u(z) \tag{3}$$

for all $z \in D$ _and_ $r = r_1, r_2$. _Then_ u _is harmonic in_ Δ _unless the equations_

$$P_s(\cosh r_1) = 1 \qquad P_s(\cosh r_2) = 1$$

have a common nonzero solution $s \in \mathbf{C}$. Here P_s denotes the usual Legendre function.

The corresponding theorem for analytic functions mixes euclidean and hyperbolic notions in an amusing fashion: condition (3) becomes

$$\int_{\Gamma_r(z)} u(\zeta)d\zeta = 0.$$

Proofs of these results can be based on the Delsarte-Lions theory of transmutation operators [2]. We prefer, instead, to make use of the Fourier analysis of SL(2, R) developed by Ehrenpreis and Mautner [4].

References

1. R. Courant and D. Hilbert. Methods of Mathematical Physics, vol. 2, Interscience, 1962.
2. J. Delsarte and J. L. Lions. Moyennes généralisées, Comment. Math. Helvet., 33 (1959), 59-69.
3. L. Ehrenpreis. Fourier analysis in several complex variables, Wiley-Interscience, 1970.
4. L. Ehrenpreis and F. I. Mautner. Some properties of the Fourier-transform on semisimple Lie groups III, Trans. Amer. Math. Soc., 90 (1959), 431-84.
5. L. Schwartz. Théorie générale des fonctions moyennes-périodiques, Ann. of Math., (2) 48 (1947), 857-929.
6. C. L. Siegel. Über einige Anwendungen diophantischer Approximationen, Abh. Preuss. Akad. Wiss. Phys.-Math. Kl. (1929), no. 1.
7. L. Zalcman. Analyticity and the Pompeiu problem, Arch. Rat. Mech. Anal., 47 (1972), 237-54.
8. L. Zalcman. Mean values and differential equations, Israel J. Math., 14 (1973), 339-52.

University of Maryland,
College Park, U. S. A.

RESEARCH PROBLEMS IN FUNCTION THEORY

W. K. HAYMAN

In September 1964 a N. A. T. O. advanced study institute was held at Imperial College, London University. Arising out of this a book of problems with the above title was collected by me and published by the Athlone Press in 1967. It is my aim in this article to report on the progress that has been made on the problems of this book in so far as I am aware of it. I have given the references that I could easily trace. However, much of the work is as yet unpublished. Some of the solutions were first announced at the present conference in Canterbury. After discussing progress on the old problems I shall describe some new problems which were contributed by members of this conference.

In addition to the above problem collection the attention of the reader should also be drawn to the recent collection of Ch. Pommerenke.

I acknowledge with gratitude the very considerable help I have had from Professor Clunie, particularly in the tracing of references.

Progress on the previous problems

The problems are numbered as in the book referred to in the introduction [P]. The notation and terminology of that book will be used throughout.

1. Meromorphic functions

1.1 This problem has now been completely settled by Drasin in a result first announcedat the Canterbury conference. An abstract appears elsewhere in these proceedings. Drasin not only constructs a function with preassigned deficiencies $\delta(a)$, but also with preassigned branching indices $\theta(a)$.

1.2 Progress on this problem has been made by Hyllengren and me. Hyllengren has shown that all values of a set E can have Valiron deficiency greater than a positive constant for a function of finite order in the plane if and only if there exists a sequence of complex numbers a_n and a $k > 0$ such that each point of E lies in infinitely many of the disks

$$\{z : |z - a_n| < e^{-e^{kn}}\}.$$

I proved that all values of any F_σ set of capacity zero can be Valiron deficiencies for an entire function of infinite order. More precisely given any two real valued functions $\phi_1(r)$ and $\phi_2(r)$ tending to infinity with r, there exists a sequence r_ν tending to infinity with ν and an entire function f(z), such that for $a \in E$

$$(1) \qquad T(r, f) > \phi_1(r) \qquad (r > r_0)$$

and

$$(2) \qquad N(r_\nu, a) < \phi_2(r_\nu)\log r_\nu \qquad (\nu > \nu_0(a)).$$

Taking e. g., $\phi_1(r) = r$, $\phi_2(r) = \log r$ we see that all values of E have Valiron deficiency 1. It remains open whether the corresponding result holds for sets of capacity zero other than F_σ sets or whether for sets of positive capacity Ahlfors' theorem [P, (1. 3)] can be improved.

1. 3 Weitsman has proved that the number of deficiencies is at most twice the order in this case.

1. 4 The answer is no, even in a very weak sense. Gol'dberg has constructed an example of an entire function for which

$$n_1(r, a) = O((\log r)^{2+\varepsilon}), \quad n_1(r, b) = O((\log r)^{2+\varepsilon}) \qquad (r \to \infty),$$

but the order is not a multiple of $\frac{1}{2}$.

1. 7 and 1. 8 A better lower bound has recently been obtained by Miles and Shea who also obtained the exact lower bound for any order λ of

$$\limsup_{r\to\infty} \frac{N(r, 0) + N(r, \infty)}{m_2(r)},$$

where

$$m_2(r) = \{\frac{1}{2\pi}\int_0^{2\pi} (\log|f(re^{i\theta})|)^2 d\theta\}^{\frac{1}{2}} .$$

Hellerstein and Williamson have solved the problem completely for entire functions with zeros on a ray.

1.9 This had already been settled in the paper of Gol'dberg quoted in [P].

1.10 and 1.11 Drasin has shown that the answer to each of these questions is no. There exists an entire function of finite order with several deficient values such that

$$\limsup_{r\to\infty} \frac{T(\sigma r)}{T(r)} = \infty$$

for any $\sigma > 1$.

1.13 The condition $1 < \lambda < \infty$ should read $2 < \lambda < \infty$. An affirmative answer to this problem, giving the exact value of $B(\lambda)$, is published elsewhere in these proceedings by Hellerstein and Shea. They show that $B(\lambda) = 1 - \{\frac{\pi^2}{e} + o(1)\}/\log \lambda$ as $\lambda \to \infty$.

1.14 This has been completely settled by Weitsman, who proved that $\Sigma\{\delta(a, f)\}^{1/3}$ does indeed converge for any meromorphic function of finite order.

1.15 This result (known as Edrei's spread conjecture) was recently proved by Baernstein by means of his function $T^*(r, \theta)$. With this important new tool Baernstein has been able to settle various other problems, which will be discussed later.

1.17 This inequality, called Paley's conjecture, has been proved by Govorov for entire functions and by Petrenko for meromorphic functions.

1.18 The conjecture has been proved by Mues if f has finite order and $ff'' \neq 0$ (instead of $ff'f'' \neq 0$). The case of infinite order seems very difficult.

1.22 A negative answer to this question is provided by the examples discussed in connection with problem 1.2. These show that the second fundamental theorem fails to hold on the sequence $r = r_\nu$.

1.23 Gol'dberg has given examples of meromorphic functions of finite order for which the deficiency is not invariant under change of origin. The case of entire functions of finite order remains open.

2. Entire functions

2.1 The first part was really settled by the examples of Arakelyan. Let E be a dense countable set in the plane every value of which is deficient. Then clearly

$$\mu(r, f-\omega) \to 0 \quad (r \to \infty)$$

for every $\omega \in E$ and so for every ω in the plane. I owe this observation to Gol'dberg.

The second part was settled by Drasin and Weitsman. The set of ω for which $m(r, \frac{1}{f-\omega}) \to \infty$ $(r \to \infty)$ must have capacity zero and an arbitrary set of capacity zero may occur.

2.2 Such a construction has been given by me.

2.4 An example has been given by Toppila of a function having different exceptional values at each of n Julia lines.

2.5 Gol'dberg has answered several of the questions posed in this problem. In particular he showed that, given any countable set A there exists E such that $A \subset E \subset \bar{A}$, where $\bar{A}$ is the closure of E. I believe that he has also done something on 2.4.

2.8 Gol'dberg has pointed out to me that instead of order λ we should read lower order μ. Otherwise the answer is negative.

2.16 Herzog and Piranian have shown that (a) is indeed possible. However, the answer to (b) is still unknown. These authors also provided an example of a univalent function in the unit disk for which the analogue of (a) holds.

2.17 Kjellberg showed in a lecture at our conference that my conjecture is false and that $.24 < A < .248$. In this problem and also in problem 2.18 the monomials az^n should be excluded.

2.21 Baker has shown that every entire function other than a linear polynomial does indeed have repulsive fixed points and these are dense in the set of non-normality of the function.

2.22 Baker has shown that if $f(z) = kze^z$, where k is a certain positive constant, then the set of non-normality does indeed occupy the whole plane. The problem of whether $f(z) = e^z$ has this property remains open.

2.25 Read 'linearly independent' for 'distinct'. No further progress seems to have been made in this problem.

2.28 A positive answer to the first part was recently provided by me, when I showed that b.v.d. functions are precisely those whose derivatives have bounded index. A negative answer to the second part was provided by Gol'dberg. Another such example is given by the sigma-function. If E is any bounded set and $\varepsilon < \frac{1}{2}$, then for $\omega \in E$ the equation $\sigma(z) = \omega$ has exactly one root in $|z - m - in| < \varepsilon$ for integers m, n with $m^2 + n^2 > r_0(\varepsilon, E)$. But $\sigma(z)$ has order 2 and so cannot be a b.v.d. function.

2.29 Such a characterisation has been provided by Tijdeman. The coefficients $f_\nu(z)$ of the differential equation

$$y^{(n)} + f_1(z)y^{(n-1)} + \ldots + f_n(z)y = 0 \tag{3}$$

are polynomials if and only if there exist fixed numbers p and q such that each solution $g(z)$ of (3) is p-valent in any disk $\{z : |z - z_0| < 1/(1 + r^q)\}$, where $r = |z_0|$.

2.30 **and** 2.31 Functions satisfying the conditions of these problems have been constructed by Barth and Schneider.

3. Subharmonic and harmonic functions

3.2 This problem has now been largely settled. If $u(x)$ is subharmonic and has finite least upper bound M, then I showed that $u(x) \to M$ $(x \to \infty)$ along almost all rays through the origin. If $u(x)$ is not bounded above, and in particular if $u(x)$ is harmonic, there always exists a path Γ such that

$$(4) \qquad u(x) \to +\infty \quad (x \to \infty \text{ along } \Gamma).$$

This is a recent result of Fuglede which completes earlier results of Talpur and myself. Fuglede uses a deep theorem about Brownian motion by Nguyen-Xuan-Loc and T. Watanabe. The path Γ is locally a Brownian motion and so highly irregular. It is an open question whether a smooth path or a polygonal path exists satisfying (4). For continuous subharmonic functions, and in particular for harmonic functions, such paths certainly exist. I am informed that Carleson has now proved the existence of a polygonal path for general subharmonic functions.

3.3 The answer is no. I have constructed an example of a function satisfying Hall's conditions and such that $u(r) > -\frac{1}{2}K$ on the whole positive axis.

3.5 Hornblower has recently shown that (3.4) can indeed be replaced by the much weaker condition

$$\int_0^1 \log^+ B(r)dr < \infty,$$

and that this is more or less best possible. However, it is still an open question as to whether condition (3.3) is sharp.

3.7 The result, generalizing Govorov's theorem to subharmonic functions in space, has been obtained by Dahlberg.

3.10 The result has been proved by Mergelyan provided the complement of E is sufficiently thin at some point of E. However, the complete question remains open. Mergelyan lectured on this topic at Nice, but I have been unable to trace a published result.

4. Polynomials

4.2 This should read $M \geq 2A$. The inequality $|a_\nu| \leq \frac{1}{2}M$ has now been proved by Saff and Sheil-Small except when n is even and $\nu = \frac{1}{2}n$.

4.5 The conjecture appears to be due to Sendov. It has been proved for $n \leq 5$ by Meir and Sharma and for general n by Rubinstein if $|z_1| = 1$.

4.9 Professor Erdős writes 'Pommerenke showed that to every $\varepsilon > 0$ and k there is an integer n_0 so that for $n > n_0$ there is a polynomial $f_n(z)$ of degree n for which $E_f^{(n)}$ has at least k components of diameter greater than $4 - \varepsilon$. One could try to estimate the number of components of $E_f^{(n)}$ having diameter greater than $1 + c$. Is it $o(n)$ $(n \to \infty)$? This seems certain, but it could be $o(n^\varepsilon)$ $(n \to \infty)$.'

4.12 An affirmative answer has been given by Elbert.

4.17 Hálasz has just proved this conjecture with $c_3 = c_4 = 1$ together with the analogous result for trigonometric polynomials.

4.20 The problem is in fact due to Erdős. It has been settled affirmatively by Saff and Sheil-Small. They have also obtained the sharp bounds for

$$\int_{-\pi}^{\pi} |f_n(e^{i\theta})|^p d\theta$$

whenever $f_n(z) = z^n + \ldots$ is a polynomial of degree n all of whose zeros lie on $|z| = 1$ and $p > 0$.

5. Functions in the unit disk

5.1 and 5.3 Substantial progress has been made by Pommerenke. He has shown that if $f(z)$ satisfies

$$\left|\frac{f'(z)}{f(z)}\right| \leq \frac{\alpha}{1 - |z|} \qquad (r_0 < |z| < 1),$$

then

$$I_1(r, f) = O((1 - r)^{-\lambda(\alpha)}) \qquad (r \to 1)$$

and hence

$$a_n = O(n^{\lambda(\alpha)}),$$

where $\lambda(\alpha) = \frac{1}{2}(\sqrt{1+4\alpha^2} - 1)$. In particular if f is weakly univalent we may take $\alpha = 2 + \varepsilon$, $\lambda(\alpha) = \frac{1}{2}(\sqrt{17} - 1) + \varepsilon = 1.562\ldots + \varepsilon$.

5.8 Turan has shown that if $f(z) = \sum_1^\infty a_n z^n$ is univalent in $|z| < 1$ with $\sum_p^{2p} |a_n|^2 \geq 1$ for some p, then Bloch's conclusion holds with $B \geq (32e^2)^{-1}$. An analogous result holds with $2p$ replaced by Cp, where C is any constant greater than 1, but not if C is allowed to tend to ∞ with p, however slowly. This latter result is due to Petruska.

5.12 The result has been proved by Chang and Yang for $n \geq 2$. The case $n = 1$ appears to be still open.

5.14 The result has been proved by Drasin if f is entire.

5.16 These results have now been proved by Baernstein if D (and so D^*) is simply-connected.

6. Schlicht and multivalent functions

6.1 The result for $n = 6$ has been proved independently by Pederson and Ozawa, and for $n = 5$ by Pederson and Schiffer.

6.2 Recently Fitzgerald proved $|a_n| \leq n\sqrt{\frac{7}{6}} = 1.08\ldots n$ for all $n \geq 2$.

6.5 In this connection Pommerenke has proved that

$$|b_n| \geq n^{-17-1}$$

can hold for infinitely many n.

6.8 Baernstein has now shown that if $\phi(R)$ is any convex function of $\log R$ then for each r $(0 < r < 1)$,

$$\int_0^{2\pi} \phi(|f(re^{i\theta})|)d\theta$$

is maximised in the class S by the Koebe function. Thus he has completely settled this classical problem.

6.9 Ruscheweyh and Sheil-Small have now proved this result known as Schoenberg's conjecture.

6.10 Pommerenke believes that this has now been proved.

6.11 A counter-example has been given by Macgregor, who has also found the largest disk in which $\lambda f + (1 - \lambda)g$ is starlike.

6.13 Pommerenke has made some progress with this.

6.14 Pommerenke has proved that for $k = 2, 3, \ldots$

$$A_n^{(k)} = O(n^{3/2 - \frac{1}{2}k - k\beta}) \ (n \to \infty)$$

for some constant $\beta > \frac{1}{4000}$.

6.13' and 6.14' Some work on these problems has been done by Noonan and Thomas.

6.15 Some improvements are due to Becker.

6.16 A counter-example to the Marx conjecture has been given by Hummel.

7. Miscellaneous

7.5 The plane case has been settled independently by Katona and Kleitman. The general case has been dealt with by Kleitman.

7.6 If the arguments of the coefficients are allowed to vary a complete answer has been given by Offord. In this case it is true with probability one that $f(z)$ assumes in every angle in the unit disk every value in the plane.

7.7 This has now been settled by Buckholz. He has found a method by which the constant in question can be calculated with any desired accuracy.

References

The numbers after the authors' names refer to the numbers of the problems in the preceding text. The letter C stands for a result published or announced elsewhere in these proceedings. The letter U stands for a hitherto unpublished result. In some cases where I have been unable to trace the reference in the time available to me I have left it blank. When I had a partial reference I have put it in.

A. Baernstein. 1.15, Proc. Lond. Math. Soc. (3) 26 (1973), 418-34; 5.16, 6.8 and 6.26, C, 11-15.

I. N. Baker. 2.21, Math. Z. 104 (1968), 252-6; 2.22, Ann. Acad. Sci. Fenn. A, 469 (1970), 11pp.

K. F. Barth and W. J. Schneider. 2.30, Proc. Amer. Math. Soc. 32 (1972), 229-32; 2.31, J. Lond. Math. Soc. (2) 4 (1972), 482-8.

J. Becker. J. reine angew. Math. 255 (1972), 23-43.

J. D. Buckholz. 7.7, Michigan Math. J. 17 (1970), 5-14.

K. Chang and L. Yang. 5.12, Sci. Sinica 14 (1965), 1262.

B. Dahlberg. 3.7, Arkiv för Math. 10 (1973), 293-309.

D. Drasin. 1.1, C, 31-41; 1.10 and 1.11, U; 5.14, Acta Math. 122 (1969), 231-63.

D. Drasin and A. Weitsman. 2.1, Indiana U. Math. J. 20 (1971), 699-715.

A. Elbert. 4.12, Stud. Sci. Math. Hungar. 1 (1966), 119-28 and ibid. 3 (1968), 299-324.

P. Erdős. 4.20, Bull. Amer. Math. Soc. (1940), 953-8.

C. Fitzgerald. 6.2, Arch. Rational Mech. Anal. 46 (1972), 356-68.

B. Fuglede. 3.2, U.

A. A. Gol'dberg. 1.4, Sib. Math. J. (1973); 1.9, D. A. N. S. S. S. R. 159 (1964), 968-70; 1.23, Translation of H. Wittich's book Neuere Untersuchungen über eindeutige analytische Funktionen into Russian, addendum §1, 264-6. See also Gol'dberg's recent book with A. Ostrovskii, (Moscow 1970, pp. 201-2). 2.4, ?; 2.5, Acta Math. Acad. Sci. Hungar. 19 (1968); 2.1, 2.8 and 2.24, Letters to the author, 1968; 2.28, Izv. Akad. Uzbek. Nauk (1972).

N. V. Govorov. 1.17, Funkz. Anal. 3 (1969), 35-40.

G. Halász. 4.17, Studia Sci. Math. Hungar. (to appear).

W. K. Hayman. Research problems in function theory (Athlone Press, 1967) [P]; 1.2 and 1.22, Ark. för Math. 10 (1972), 163-72; 2.2, Proc. Cam. Phil. Soc. 66 (1969), 301-15; 2.28, Pacific J. Math. 44 (1973), 117-37; 3.2 and 3.3, these results will be contained in a forthcoming monograph on subharmonic functions, chapters 4 and 7 respectively.

S. Hellerstein and D. Shea. 1.13, C, 81-7.

S. Hellerstein and J. Williamson. 1.7, J. d. Analyse Math. 22 (1969), 233-67.

F. Herzog and G. Piranian. 2.16, Proceedings of Symposia in Pure Mathematics, 11, Entire functions and related parts of analysis, Amer. Math. Soc. (1968), 240-3.

R. M. Hornblower. 3.5, Proc. Lond. Math. Soc. (3) 23 (1971), 371-84 and Ann. Polon. Math. 26 (1972), 135-46.

J. A. Hummel. 6.16, Michigan Math. J. 19 (1972), 257-66.

A. Hyllengren. 1.2, Acta Math. 124 (1970), 1-8.

Gy. Katona. 7.5, Studia Sci. Math. Hungar. 1 (1966), 59-63.

B. Kjellberg. 2.17, U.

D. Kleitman. 7.5, Math. Z. 90 (1965), 251-9.

T. H. Macgregor. 6.11, J. Lond. Math. Soc. 44 (1969), 210-2.

A. Meir and A. Sharma. 4.5, Pacific J. Math. 31 (1969), 459-67.

S. M. Mergelyan. 3.10, ?.

J. Miles and D. F. Shea. 1.7 and 1.8, Quart. J. Math. (2), 24 (1973), 377-83.

E. Mues. 1.18, Math. Z. 119 (1971), 11-20.

Nguyen-Xuan-Loc and T. Watanabe. 3.2, Z. f. Wahrscheinlichkeitstheorie u. verw. Geb. 21 (1972), 167-78.

J. W. Noonan and D. K. Thomas. 6.13' and 6.14', Proc. London Math. Soc. (3) 25 (1972), 503-24.

A. C. Offord. 7.6, Studia Math. 41 (1972), 71-106.

M. Ozawa. 6.1, Kodai Math. Sem. Reports 21 (1969), 97-128, and ibid. 129-32.

R. N. Pederson. 6.1, Arch. Rational. Mech. Anal. 31 (1968), 331-51.

R. N. Pederson and M. Schiffer. 6.1, Arch. Rational Mech. Anal. 45 (1972), 161-93.

P. Petrenko. 1.17, Izv. Akad. Nauk S.S.S.R. 33 (1969), 444-54.
G. Petruska. 5.8, Ann. Univ. Sci. Budapest. Eötvös Sect. Math. 12 (1969), 39-42.
Ch. Pommerenke. 4.9, Michigan Math. J. 8 (1961), 97-125; 5.1 and 5.3, J. Lond. Math. Soc. (2) 5 (1972), 624-8; 6.5, Inventiones math. 3 (1967), 1-15; 6.13, Invent. Math. 3 (1967), 1-15; 6.14, Mathematika 14 (1967), 108-12; Problems in complex function theory, Bull. Lond. Math. Soc. 4 (1972), 354-66.
Z. Rubinstein. 4.5, Pacific J. Math. 26 (1968), 159-61.
S. Ruscheweyh and T. B. Sheil-Small. 6.9, Comm. Math. Helv. 48 (1973), 119-35.
E. Saff and T. B. Sheil-Small. 4.2 and 4.20, U.
M. N. M. Talpur. 3.2, On the sets where a subharmonic function is large. Thesis (London, 1967).
R. Tijdeman. 2.29, On the distribution of values of certain functions. Thesis (Amsterdam, 1969).
S. Toppila. 2.4, Acta Soc. Sci. Fenn. A, 456 (1970), Theorem 9.
P. Turan. 5.8, Math. Lapok. 17 (1966), p. 224.
A. Weitsman. 1.3, Acta Math. 123 (1969), 115-39 also C, 133-5; 1.14, Acta Math. 128 (1972), 41-52.

NEW PROBLEMS

The numbers will run consecutively with the corresponding numbers in [P]. The notation will also be similar to that of [P].

1. Meromorphic functions

1.24 If f is meromorphic in the plane can $n(r, a)$ be compared in general with its average value

$$A(r) = \frac{1}{\pi} \iint_{|z|<r} \frac{|f'(z)|^2}{(1 + |f(z)|^2)^2} \, dxdy$$

in the same sort of way that $N(r, a)$ can be compared with $T(r)$? In particular is it true that

$$n(r, a) \sim A(r)$$

as $r \to \infty$ outside an exceptional set of r independent of a and possibly an exceptional set of a? (Compare problem 1.16.)

1.25 In the opposite direction to the previous problem does there exist a meromorphic function such that for every pair of distinct values a, b we have

$$\limsup_{r\to\infty} \frac{n(r, a)}{n(r, b)} = \infty, \qquad \liminf_{r\to\infty} \frac{n(r, a)}{n(r, b)} = 0.$$

Note, of course, that either of the above limits for <u>all</u> distinct a, b implies the other.

(Compare the result (1.3) quoted in problem 1.2, which shows that this certainly cannot occur for the N-function.)

The above question can also be asked for entire functions.

(P. Erdős)

1.26 The analogue of problem 1.7 may be asked for meromorphic functions. The proposers conjecture that in this case

$$\sum_a \delta(a, f) \le \max\{\Lambda_1(\lambda), \Lambda_2(\lambda)\},$$

where for $\lambda \ge 1$, $q = [2\lambda]$ we have

$$\Lambda_1(\lambda) = 2 - \frac{2 \sin \frac{1}{2}\pi(2\lambda - q)}{q + 2 \sin \frac{1}{2}\pi(2\lambda - q)},$$

$$\Lambda_2(\lambda) = 2 - \frac{2 \cos \frac{1}{2}\pi(2\lambda - q)}{q + 1}.$$

In an abstract of these proceedings the second author shows that this result would be sharp. The correct bound is known for $0 \le \lambda \le 1$.

(D. Drasin and A. Weitsman)

1.27 Let E be the set of a for which

$$m(r, a) \to \infty \quad (r \to \infty).$$

How large can the set of E be if

(a) f is entire and of order $\frac{1}{2}$ mean type,

(b) f is meromorphic of order λ, where $0 \le \lambda \le \frac{1}{2}$.

The proposers settled this problem in all other cases. (See the above progress report on problem 2.1.)

(D. Drasin and A. Weitsman)

1.28 Are there upper bounds of any kind on the set of asymptotic values of a meromorphic function of finite order?

(D. Drasin and A. Weitsman)

1.29 Under what circumstances does there exist a meromorphic function f(z) of finite order ρ with preassigned deficiencies $\delta_n = \delta(a_n, f)$ at a preassigned sequence of complex numbers? Weitsman has solved problem 1.14 (see earlier progress report) by showing that it is necessary that

$$(1.5) \quad \sum \delta_n^{1/3} < \infty,$$

but the bound on the sum of the series depends on the biggest term δ_1. On the other hand I showed (M. F., p. 98) that the condition

$$(1.6) \quad \sum \delta_n^{1/3} \le A$$

with $A = 9^{-1/3}$ is sufficient to yield a meromorphic function of order 1 mean type such that $\delta(a_n, f) \geq \delta_n$. Possibly (1.6) with a constant $A = A_1(\rho)$ is sufficient and with a larger constant $A = A_2(\rho)$ is necessary. If ρ is allowed to be arbitrary but finite the problem may be a little easier.

2. Entire functions

Minimum modulus problems

Let $f(z)$ be an entire function of order λ and lower order μ, and let

$$L(r) = \min_{|z|=r} |f(z)|, \quad M(r) = \max_{|z|=r} |f(z)|.$$

It is a classical result that

$$\limsup_{r\to\infty} \frac{\log L(r)}{\log M(r)} \geq C(\mu).$$

Here $C(\mu) = \cos \pi\mu$ for $0 \leq \mu \leq 1$, and [Hayman]

$$-2.19 \log \mu < C(\mu) < -.09 \log \mu$$

when μ is large. We refer the reader to Barry for a general account of the situation. For functions of infinite order the analogous result is

$$\limsup_{r\to\infty} \frac{\log L(r)}{\log M(r) \log \log \log M(r)} \geq C_\infty,$$

where

$$-2.19 < C_\infty < -.09.$$

There are a number of open questions.

2.33 Is it possible to obtain the exact value of C_∞ or the asymptotic behaviour of $\frac{C(\mu)}{\log \mu}$ as $\mu \to \infty$? The question seems related to the number of zeros a function can have in a small disk centred on a point of $|z| = r$. [See Hayman, loc. cit.]

2.34 Is it possible to say something more precise about $C(\mu)$ when μ is just greater than 1? In particular is it true that $C(\mu) = -1$ for such μ or, alternatively, is $C(\mu)$ a strictly decreasing function of μ?

2.35 If Γ is a continuum that recedes to ∞ it is known [Hayman, loc. cit.] that

$$\limsup_{\substack{z \to \infty \\ z \in \Gamma}} \frac{\log |f(z)|}{\log M(|z|)} \geq -A,$$

where A is an absolute constant. Is it true that $A = 1$? This is certainly the case if Γ is a ray through the origin [Beurling]. If $A > 1$ is it possible to obtain a good numerical estimate for A?

Other questions relate to the case $0 \leq \mu \leq 1$, where much more is known.

2.36 Suppose that $0 < \lambda < \alpha \leq 1$, where λ is the order. Let E_α be the set of r for which

$$\log L(r) > \cos \pi\alpha \log M(r).$$

Besicovitch showed that the upper density of E_α is at least $1 - \lambda/\alpha$, and Barry proved the stronger result that the same is true of the lower logarithmic density of E_α. Examples given by me show that Barry's theorem is sharp; in these examples the logarithmic density exists, but the upper density is larger. This suggests that Besicovitch's theorem may be sharpened.

2.37 Let r_n be a sequence of Pólya peaks [Edrei] of order λ. Then Edrei [loc. cit.] showed that there exists $K = K(\alpha, \lambda)$ such that

$$\log L(r) > \cos \pi\alpha \log M(r)$$

for some value r in the interval $r_n \leq r \leq Kr_n$ and n sufficiently large. Is $K(\alpha, \lambda)$ independent of α for fixed λ? Can it be taken arbitrarily near 1?

(D. Drasin and A. Weitsman)

2.38 It was shown by Kjellberg that if $0 < \alpha < 1$ and

$$\log L(r) < \cos \pi\alpha \log M(r) + O(1) \qquad (r \to \infty),$$

then

$$\lim_{r\to\infty} \frac{\log M(r)}{r^{\alpha}} = \beta,$$

where $0 < \beta \leq \infty$. If $\alpha = 1$ it was recently shown by me that, unless $f(z) = Ae^{Bz}$, the corresponding result holds with $\beta = \infty$. Examples constructed by me showed that

$$L(r) M(r) \to 0 \qquad (r \to \infty)$$

can occur for a function of order $1 + \varepsilon$ for any $\varepsilon > 0$. The case of functions of order 1 and maximal type remains open.

2.39 We can also compare $L(r)$ with the characteristic $T(r)$. We have

$$\limsup_{r\to\infty} \frac{\log L(r)}{T(r)} \geq D(\mu)$$

and ask for the best constant $D(\mu)$. In view of Petrenko's solution of problem 1.17 we certainly have

$$D(\mu) \geq -\pi\mu \qquad (1 \leq \mu < \infty).$$

Also Essén and Shea show that $D(\mu) \leq \frac{\pi\mu}{1 + |\sin \pi\mu|}$ for $1 < \mu < \frac{3}{2}$ and $D(\mu) \leq \frac{-\pi\mu}{2}$ for $\frac{3}{2} < \mu < \infty$. Further it follows from results of Valiron, Edrei and Fuchs that

$$D(\mu) = \begin{cases} \pi\mu \cot \pi\mu & (0 \leq \mu < \frac{1}{2}) \\ \pi\mu \cos \pi\mu & (\frac{1}{2} \leq \mu < 1). \end{cases}$$

(D. Shea)

2.40 Let $f(z)$ be a non-constant entire function and assume that for some c the plane measure of the set $E(c)$ where $|f(z)| > c$ is finite. What is the minimum growth rate of $f(z)$? Hayman conjectures that

$$\int^{\infty} \frac{r\,dr}{\log\log M(r)} < \infty$$

is true and best possible.

If $E(c)$ has finite measure is the same true of $E(c')$ for $c' < c$?

(P. Erdős)

2.41 Suppose that $f(z)$ has finite order and P is a rectifiable path on which $f(z) \to \infty$. Let $l(r)$ be the length of P in $|z| < r$. Find such a path for which $l(r)$ grows as slowly as possible and estimate $l(r)$ in terms of $M(r)$. If $f(z)$ has zero order, or more generally finite order, can a path be found for which $l(r) = O(r)$ $(r \to \infty)$? If $\log M(r) = O(\log^2 r)$ $(r \to \infty)$, but under no weaker growth condition it is known that we may choose a ray through the origin for P [Hayman, Piranian].

If $f(z)$ has a finite asymptotic value a the corresponding question may be asked for paths on which $f(z) \to a$.

(P. Erdős)

2.42 Let $f(z)$ be an entire function (of sufficiently high order) with n (≥ 2) different asymptotic values a_k. Suppose that γ_k is a path such that

$$f(z) \to a_k (z \to \infty,\ z \in \gamma_k).$$

Let $n_1(r, a_k)$ be the number of zeros of $f(z) - a_k$ on γ_k and in $|z| \leq r$. Can we find a function $f(z)$ such that

$$\frac{n_1(r, a_k)}{n(r, a_k)} \to b_k > 0$$

as $r \to \infty$, for $k = 1, 2, \ldots, n$? Can we take $b_k = 1$?

(J. Winkler)

2.43 Let $f(z)$ be a transcendental entire function which permutes the integers, i. e. gives a 1-1 mapping of the integers onto themselves. Is it true that $f(z)$ is at least of order 1, type π? We can also ask the corresponding question for a function permuting the positive integers with the same conjectured answer. Note that $f(z) = z + \sin \pi z$ satisfies both conditions and is of order 1, type π.

If $f(z)$ assumes integer values on the positive integers, then Hardy and Pólya proved [cf. Whittaker, Theorem 11, p. 55] that $f(z)$ is at least of order 1, type log 2; and if $f(z)$ assumes integer values on all the integers, then $f(z)$ is at least of order 1, type $\frac{3+\sqrt{5}}{2}$ [Buck].

(L. Rubel)

2.44 For $f(z)$ entire of order λ and non-constant let $\nu(r)$ be the number of points on $|z| = r$ where $|f(z)| = 1$. Is it true that

$$\limsup_{r \to \infty} \frac{\log \nu(r)}{\log r} = \lambda \ ?$$

If one replaces $\nu(r)$ by the number of points on $|z| = r$ where $f(z)$ is real then the corresponding upper limit is always equal to λ [Hellerstein and Korevaar].

(J. Korevaar)

2.45 Let $J_0(z)$ be the Bessel function of order zero. Is it true that the equation $J_0(z) = 1$ has at most one solution on each ray from the origin? An affirmative answer would show that the exceptional set in a theorem of Delsarte is, in fact, void. Asymptotic estimates show that there can be at most a finite number of solutions on any ray and yield even stronger information.

(L. Zalcman)

2.46 Let $\{f_\alpha(z)\}$ be a family of entire functions and assume that for every z_0 there are only denumerably many distinct values of $f_\alpha(z_0)$. Then if $c = 2^{\aleph_0} > \aleph_1$ the family $\{f_\alpha(z)\}$ is itself denumerable. The above result was proved by Erdős. If $c = \aleph_1$ he constructed a counter-example.

Suppose now that m is an infinite cardonal, $\aleph_0 < m < c$. Assume that for every z_0 there are at most m distinct values $f_\alpha(z_0)$. Does it then follow that the family $\{f_\alpha(z)\}$ has at most power m? If $m^+ < c$, where m^+ is the successor of m, it is easy to see that the answer is yes. However, if $c = m^+$ the counter-example fails. Possibly the question is undecidable.

(P. Erdős)

3. Subharmonic and harmonic functions

We can ask the analogues of 2.31 to 2.42 for subharmonic functions in the plane. The results and even the proofs are likely to be similar as they are for the quoted theorems. For subharmonic functions in R^m with $m \geq 3$ not all problems extend in a sensible way. Thus if x denotes $(x_1, x_2, x_3, \ldots, x_m)$ and i is the unit vector along the x_1 axis, then the function

$$u(x) = -\int_0^\infty \frac{dt}{(1+t)|x+ti|^{m-2}}$$

is subharmonic in R^m, but $L(r) = -\infty$ for every r. However, the analogues of 2.40 to 2.42 still make sense in R^m though the problems are likely to be harder.

Here are some further problems.

3.11 If $u(x)$ is a homogeneous harmonic polynomial of degree n in R^m what are the upper and lower bounds of

$$-\frac{A(r, u)}{B(r, u)},$$

where $A(r, u) = \inf_{|x|=r} u(x)$, $B(r, u) = \sup_{|x|=r} u(x)$?

If n is odd it is evident that $A(r, u) = -B(r, u)$, but if $u(x) = x_1^2 + x_2^2 - 2x_3^2$ in R^3, then $B(r, u) = r^2$, $A(r, u) = -2r^2$. For transcendental harmonic functions such that $u(0) = 0$ we can prove that

$$-A(r, u) \leq \frac{(R+r)R^{m-2}}{(R-r)^{m-1}} B(r, u) \qquad (0 < r < R),$$

by Poisson's formula and this leads to

$$(3.5) \quad -A(r, u) < B(r) \{\log B(r)\}^{m-1+\varepsilon}$$

outside a set of r of finite logarithmic measure. However, (3.5) is unlikely to be sharp. We note that for $m = 2$ it follows from a classical result of Wiman that, for any harmonic function u,

$$A(r) \sim -B(r)$$

as $r \to \infty$ outside a set of finite logarithmic measure.

3.12 Consider a domain of infinite connectivity in R^3 whose complement E lies in the plane $P : x_3 = 0$. Suppose further than any disk of radius $R > 0$ in P contains a subset of E having area at least ε, where ε, R are fixed positive constants. If u is positive and harmonic in D, continuous in R^3 and zero on E is it true that

$$u = cx_3 + \phi(x) \qquad (x_3 > 0),$$

where c is a constant and $\phi(x)$ is uniformly bounded? One can also ask the analogue of this result for R^m when $m > 3$. It is true in R^2 (but the author cannot remember who proved it!).

(B. Kjellberg)

3.13 Let $u(x)$ be subharmonic in R^m. One can define the quantities $n(r, 0)$, $N(r, 0)$, $T(r)$ as in Nevanlinna theory in the plane, taking the analogue of the case $u(z) = \log|f(z)|$, where $f(z)$ is an entire function [Hayman]. Define

$$\delta(u) = 1 - \limsup_{r\to\infty} \frac{N(r, u)}{T(r)}$$

If the order λ of u is less than 1 it is possible to obtain the sharp upper bound for $\delta(u)$ in terms of λ and m. The bound is attained when $u(x)$ has all its mass on a ray. [This will appear in chapter IV of a book on subharmonic functions to be published by me.] One can ask the corresponding question for $\lambda > 1$.

One can ask whether a lower bound $A(\lambda)$ can be obtained for

$\delta(u)$ if $\lambda > 1$ and all the mass of $u(x)$ lies on a ray or more generally on some suitable lower dimensional subspace S of R^m and $\lambda > \lambda_0(S)$. For fixed S we may conjecture by analogy with the case $m = 2$, that $A(\lambda) \to 1 (\lambda \to \infty)$. This is proved elsewhere in these proceedings in the case $m = 2$ by Hellerstein and Shea.

(D. Shea)

3.14 Let there be given an integrable function F on the unit circle and a point z_0 in the unit disk. The problem is to maximize $u(z_0)$, where u runs through all functions which are subharmonic in $|z| < 1$, equal to F on $|z| = 1$ and which satisfy

$$\inf_{\theta} u(re^{i\theta}) \leq 0, \qquad 0 < r < 1.$$

(A. Baernstein)

3.15 Let D be a doubly connected domain with boundary curves α and β and let z_0, z_1 be points of D. Let A, B be given real numbers. The problem is to maximize $u(z_0)$, where u runs through all functions which are subharmonic in D, take the values A and B on α and β respectively and are nonpositive on some curve connecting z_1 to α.

(A. Baernstein)

4. Polynomials

4.22 We assume the terminology of problem 4.7. If $|z_i| \leq 1$, Clunie and Netanyahu showed that a path exists joining the origin to $|z| = 1$ in $E_f^{(n)}$. What is the shortest length $L_f^{(n)}$ of such a path? Presumably $L_f^{(n)}$ can tend to infinity with n, but not too fast.

(P. Erdős)

4.23 Some of the problems 4.7 to 4.12 extend naturally to space of higher dimensions. Let x_i be a set of n points in R^m and let $E_n^{(m)}$ be the set of points for which

$$\prod_{i=1}^{n} |x - x_i| \leq 1.$$

When is the maximum volume of $E_n^{(m)}$ attained and how large can it be? Piranian observed that the ball is not extreme for $m = 3$, $n = 2$. If $E_n^{(m)}$ is connected, can it be covered by a ball of radius 2? For $m = 2$ this was proved by Pommerenke.

(P. Erdős)

4.24 Let

$$P(z) = \sum_0^n a_k z^k$$

be a self-inversive polynomial, i.e. if ξ is a zero of $P(\xi)$ with multiplicity m, then $1/\bar{\xi}$ is also a zero with multiplicity m. Is it true that $w = P(z)$ maps $|z| < 1$ onto a domain containing a disk of radius $A = \max_{0 \le k \le n} |a_k|$?

(T. Sheil-Small)

5. Functions in the unit disk

Spaces of analytic functions

5.22 Let H^p be the space of functions $f(z) = \sum_0^\infty a_n z^n$ regular in $|z| < 1$ and such that

$$\int_0^{2\pi} |f(re^{i\theta})|^p d\theta$$

remains bounded as $r \to 1$. We define H^∞ to be the class of bounded functions in $|z| < 1$. For $0 < p \le 1$ describe the coefficient multipliers from H^p to H^p. That is, for each such p describe the sequences λ_n such that

$$\sum \lambda_n a_n z^n \in H^p \quad \text{whenever} \quad \sum a_n z^n \in H^p.$$

(P. L. Duren)

5.23 Describe similarly the coefficient multipliers from S to S, where S is the class of functions $\sum_1^\infty a_n z^n$ univalent in $|z| < 1$ either

(a) with the normalisation $a_1 = 1$, or (b) generally.

(c) What are the multipliers of the space of close-to-convex functions into itself?

(d) What are the multipliers of S into the class C of convex functions?

(e) What are the multipliers from the class N of functions of bounded characteristic into itself? The analogous problem for the class N^+ may be more tractable.

Ruscheweyh and Sheil-Small in solving problem 6.9 have shown that (λ_n) is a multiplier sequence from C into itself if and only if $\Sigma \lambda_n z^n \in C$. In general one can obtain only some sufficient conditions. Thus in most cases $f(z) = \Sigma a_n z^n$ belongs to a class A if a_n is sufficiently small and conversely if $f \in A$, then a_n cannot be too big. For example $\sum_2^\infty n|a_n| \leq 1$ is a sufficient condition for $f(z) \in S$, and $|a_n| \leq n\sqrt{7/6}$ is a necessary condition [Fitzgerald]. Similarly if $\sum_1^\infty |a_n| < \infty$, then $f(z)$ is continuous in $|z| \leq 1$ and so belongs to H^p for every $p > 0$ and to N, whilst if $f \in N$, then $|a_n| \leq \exp(cn^{\frac{1}{2}})$ for some constant c. Again if $f(z)$ belongs to one of the above classes then so does $\frac{1}{t}f(tz)$ for $0 < t < 1$, so that the sequence (t^{n-1}) is a multiplier sequence. In other cases negative results are known. Thus Frostman showed that (n) is not a multiplier sequence from N to N and I showed that $(\frac{1}{n+1})$ is not such a sequence either.

(P. L. Duren except for (e) which is due to A. Shields)

5.24 Is the intersection of two finitely generated ideals in H^∞ finitely generated?

(L. Rubel)

5.25 Let W^+ be the Banach algebra of absolutely convergent power series $f(z) = \sum_0^\infty a_n z^n$ with $\|f\| = \sum_0^\infty |a_n|$.

Which functions generate W^+? More precisely for which functions f is it true that the polynomials in f are dense in W^+? It is clear that a necessary condition is that f be schlicht in the closed disk $|z| \leq 1$. Newman has shown that if in addition $f' \in H^1$ then f generates W^+. Hedberg and Lisin have shown (independently) that if f is schlicht and $\Sigma n|a_n|^2(\log n)^{1+\varepsilon} < \infty$ for some $\varepsilon > 0$, then f generates W^+. Neither Newman's condition nor Hedberg-Lisin's condition implies the other. Is schlichtness enough?

(L. Zalcman)

5.26 Let B_2 be the Bergman space of square integrable functions in the unit disk, i.e. those functions $f(z) = \sum_0^\infty a_n z^n$ for which $\sum_1^\infty n^{-1} |a_n|^2 < \infty$. A subspace S is said to be invariant if $zf \in S$ whenever $f \in S$. What are the invariant subspaces of B_2? The corresponding problem for H^2 was solved by Beurling long ago and uses the inner-outer factorisation of H^p functions, a tool unavailable in the present context.

(L. Zalcman)

5.27 **The corona conjecture.** Let D be an arbitrary domain in the plane that supports non-constant bounded analytic functions. Suppose that $f_1(z), \ldots, f_n(z)$ are bounded and analytic in D and satisfy

$$\sum_{\nu=1}^{n} |f_\nu(z)| \geq \delta > 0$$

in D. Can one find bounded analytic functions $g_\nu(z)$ in D such that

$$\sum_{\nu=1}^{n} f_\nu(z) g_\nu(z) \equiv 1$$

in D?

When D is a disk Carleson proved that the answer is yes and the result extends to finitely connected domains. The result is also known to be true for certain infinitely connected domains [Behrens, Gamelin], but false for general Riemann surfaces of infinite genus [Cole]. Presumably the answer for the general plane domain is negative. Proofs of all positive results depend on Carleson's theorem.

(L. Zalcman)

5.28 Let f be continuous in $D : |z| \leq 1$ and regular in $|z| < 1$. Let

$$\omega(f, \delta) = \sup |f(z) - f(\omega)|, \text{ for } |z - \omega| \leq \delta \text{ and } z, \omega \in D,$$
$$\tilde{\omega}(f, \delta) = \sup |f(z) - f(\omega)|, \text{ for } |z - \omega| \leq \delta \text{ and } |z| = |\omega| = 1.$$

Is it true that

$$\lim_{\delta \to 0} \frac{\omega(f, \delta)}{\tilde{\omega}(f, \delta)} = 1 \ ?$$

The proposers with B. A. Taylor have shown that there is an absolute constant C such that

$$\omega(f, \delta) \leq C\tilde{\omega}(f, \delta),$$

but that one may not take $C = 1$.

(L. Rubel, A. Shields)

5.29 Let E be a G_δ of measure zero on $|z| = 1$. Then does there exist an $f \in H^\infty$, $f \neq 0$ such that $f = 0$ on E and every point of the unit circle is a Fatou point of f?

(L. Rubel)

Bloch functions

5.30 Let $\mathcal{B}$ be the space of Bloch functions, that is the space of functions regular in $|z| < 1$ with

$$\|f\|_{\mathcal{B}} = |f(0)| + \sup_{|z|<1} (1 - |z|^2)|f'(z)| < \infty.$$

Let $\mathcal{B}_S$ be the space of functions of the form

(5.11) $f(z) = \log g'(z), \quad g \in S$

(where S is the class of problem 5.23a). Let $\mathcal{B}_Q$ be the space of functions $g \in S$ that have a quasi-conformal extension to the closed plane. (Cf. Anderson, Clunie and Pommerenke.)

(a) Is $\mathcal{B}_S$ connected in the norm topology?

(b) Is $\mathcal{B}_Q$ dense in $\mathcal{B}_S$ in the norm topology?

(L. Bers)

5.31 It was shown be Becker that

$$\{f : \|f\|_{\mathcal{B}} < 1\} \subset \mathcal{B}_Q .$$

Is the radius 1 best possible? Is it true that

$$f \in \mathcal{B}_S, \quad \limsup_{|z| \to 1} (1 - |z|^2)|f'(z)| < 1 \Rightarrow f \in \mathcal{B}_Q.$$

(Ch. Pommerenke)

5.32 Suppose that $f_n \in \mathcal{B}_S$. What does $\|f_n - f\|_{\mathcal{B}} \to 0$ $(n \to \infty)$ mean geometrically for the functions g_n related to f_n by (5.11)?

(Ch. Pommerenke)

5.33 Let T be a regular triangular lattice in the plane. Let f(z) map $|z| < 1$ onto the universal covering surface over the complement of T. Is it true that the coefficients a_n of f tend to 0 as $n \to \infty$?

(Ch. Pommerenke)

5.34 It was proved by R. Hall that every Bloch function has (possibly infinite) angular limits on an uncountably dense subset of $|z| = 1$. Do there always exist angular limits on a set of positive measure relative to some fixed Hausdorff measure, such as logarithmic measure for example.

(J. E. McMillan, Ch. Pommerenke)

5.35 Let Γ be any discontinuous group of Moebius transformations of the unit disk. Does there always exist a meromorphic function automorphic with respect to Γ and normal, i.e. such that

$$(1 - |z|^2) \frac{|f'(z)|}{1 + |f(z)|^2}$$

is bounded in $|z| < 1$?

(Ch. Pommerenke)

Power series with gaps

5.36 Let (n_k) be a sequence of positive integers such that

$$(5.12) \quad n_{k+1} > \lambda n_k, \quad \text{where } \lambda > 1,$$

and suppose that

$$(5.13) \quad f(z) = \sum_0^{\infty} a_k z^{n_k}$$

is regular in $|z| < 1$. Is it true that if $\sum_{0}^{\infty} |a_k| = \infty$, then $f(z)$ assumes every finite value

(a) at least once;

(b) infinitely often;

(c) in every angle $\alpha < \arg z < \beta$ of $|z| < 1$?

[Cf. G. Weiss and M. Weiss.]

(J. P. Kahane)

5.37 Suppose $f(z)$ is a function as in (5.13) and let

$$\rho = \limsup_{r \to 1} \frac{\log \log M(r)}{-\log(1 - r)},$$

where $M(r)$ is the maximum modulus of $f(z)$ on $|z| = r$. We do not now assume (5.12), but let $N^0(t)$ be the number of n_k not greater than t. If $N(r, a)$ is the function of Nevanlinna theory (see [P], p. 1) it is known that

$$\limsup_{r \to 1} \frac{N(r, 0)}{\log M(r)} = 1$$

provided that either

(i) $\rho > 0$ and

$$\liminf_{k \to \infty} \frac{\log(n_{k+1} - n_k)}{\log n_k} > \tfrac{1}{2};$$

(this is implicit in Wiman, cf. [Sunyer y Balaguer]) or

(ii) $\rho > \frac{1 - \beta}{\beta}$ and $N^0(t) = O(t^{1-\beta})$ $(t \in \infty)$, where $0 < \beta < 1$. [Sons]

If $0 < \rho < \frac{1 - \beta}{\beta}$ with $N^0(t) = O(t^{1-\beta})$ $(t \to \infty)$ we ask (a), (b) and (c) of the preceding problem, at least for those cases not covered above. In particular we may consider the cases $n_k = [k^\alpha]$, where $1 < \alpha \leq \frac{3}{2}$.

(L. Sons)

6. Schlicht and multivalent functions

6.27 Let

$$g(z) = z + \delta_0 + \delta_1 z^{-1} + \ldots$$

be univalent in $|z| > 1$. Is it true that for each $\varepsilon > 0$ we have

$$n|\delta_n| = O(n^{\varepsilon}). \quad \max_{0<|\nu-n|<\frac{n}{2}} (\nu|b_\nu| + 1) \quad (n \to \infty)?$$

This is suggested by some results of Clunie and Pommerenke.

(J. Clunie, Ch. Pommerenke)

6.28 Let $f(z) = z + \sum_2^{\infty} a_n z^n$ belong to S and let $P(z) = \sum_0^n b_k z^k$ be a polynomial of degree at most n. Is it true that

$$\max_{|z|=1} |P(z) * f(z)| \leq n \max_{|z|=1} |P(z)| \ ?$$

Here $P * f = \sum_0^n a_k b_k z^k$. The above result would imply Rogosinski's generalized Bieberbach conjecture but is weaker than Robertson's conjecture [Sheil-Small].

(T. Sheil-Small)

6.29 With the above notation $f(z) \in S$ if and only if for each pair of numbers ξ_1, ξ_2 satisfying $|\xi_1| \leq 1$, $|\xi_2| \leq 1$ we have

$$f(z) * \frac{z}{(1 - \xi_1 z)(1 - \xi_2 z)} \neq 0 \quad (0 < |z| < 1).$$

On the other hand it is true that if $F(z) * f(z) \neq 0$ $(0 < |z| < 1)$ for all $f \in S$ then $F(z)$ is starlike. What is the complete class of starlike functions having this property? $F(z) = z + z^n/n$ has the property if and only if Bieberbach's conjecture holds.

(T. Sheil-Small)

6.30 Let $f \in S$. Baernstein has shown that

$$\int_0^{2\pi} |f(re^{i\theta})|^p d\theta \leq \int_0^{2\pi} |k(re^{i\theta})|^p d\theta, \quad 0 < r < 1, \quad 0 < p < \infty,$$

where $k(z)$ is the Koebe function.

Does the corresponding inequality hold for integral means of the derivatives at least for certain values of p? The best we can hope for is that it holds for $p \geq \frac{1}{3}$, because $k'(z) \in H^p$ for $p < \frac{1}{3}$ and there exist functions $f(z) \in S$ for which $f'(z)$ does not belong to any H^p. (An example is due to Lohwater, Piranian and Rudin). For close-to-convex functions it was proved by Macgregor that the result holds for $p \geq 1$ and in fact the corresponding inequality holds for derivatives of all orders.

(A. Baernstein)

6.31 Duren has shown that if $f(z) = \sum_{0}^{\infty} a_n z^n \in S$ and if

$$(1 - r)^2 f(r) = \lambda + O\{(1 - r)^{\delta}\}, \quad \text{as } r \to 1-$$

for some $\lambda \neq 0$ and $\delta > 0$, then

$$\frac{a_n}{n} = \lambda + O\left(\frac{1}{\log n}\right) \quad \text{as } n \to \infty.$$

To what extent can this estimate be improved?

(P. Duren)

7. Miscellaneous

Quasiconformal mappings

7.9 Suppose that f is a plane K-quasiconformal (K-q. c.) mapping of the unit disk D onto itself. Show that there exists a finite constant $b = b(K)$ such that

$$m(f(E)) \leq b\{m(E)\}^{1/K}$$

for each measurable set $E \subset D$. Here m denotes plane Lebesgue measure. Such an inequality is known [Gehring and Reich] with the exponent $\frac{1}{K}$ replaced by a constant $a = a(K)$.

(F. W. Gehring)

7.10 It was proved by Boyarskii that the partial derivatives of a plane K-q. c. mapping are locally L^p-integrable for $2 \leq p < 2 + c$, where $c = c(K) > 0$. Show that this is true with $c = \frac{2}{K - 1}$.

The example $f(z) = |z|^{\frac{1}{K}-1} . z$ shows that such a result would be sharp.

(F. W. Gehring)

7.11 Show that each q. c. mapping of R^n onto R^n has a q. c. extension to R^{n+1}. This has been established by Ahlfors when $n = 2$ and by Carleson when $n = 3$.

(F. W. Gehring)

7.12 Suppose that f is an n dimensional K-quasiregular function. Show that the partial derivatives of f are locally L^p-integrable for $n \le p \le n + c$, where $c = c(K, n) > 0$. This was shown by the proposer to be true if f is 1-1.

(F. W. Gehring)

Riemann surfaces and conformal mapping

7.13 One part of Nevanlinna theory is devoted to the following problem. How does the geometric structure of a simply connected Riemann covering surface of the sphere influence the value distribution of the meromorphic function generating the surface? It is suggested that one consider among the constituent pieces of such surfaces also halfsheets which have an infinite number of branch points on their boundary.

A typical example is the covering surface generated by $e^z - z$. It contains a right half plane with a second order branch point at $2\pi in$ for each integer n.

(F. Huckeman)

7.14 **Boundary values of Cauchy integrals.** Let γ be a C^1 curve in the plane and let f be a continuous function on γ. Put

$$F(z) = \int_\gamma \frac{f(t)dt}{t - z} \qquad (z \notin \gamma)$$

Does F have non-tangential boundary values a. e. on γ? This is a very old question which has been studied quite extensively, especially by Russian mathematicians. It is known that the answer is yes if slightly greater smoothness is assumed for γ or f (see e. g. [Havin] for an outstanding recent contribution to this subject).

Suppose now that γ is a Jordan curve and let ϕ be a conformal map from the unit disk onto the inside of γ. Does $F \circ \phi$ belong to H^p for some $p < 1$ or perhaps to the class N of functions with bounded Nevanlinna characteristic?

(A. Baernstein)

7.15 Let D be a domain in the extended complex plane. A finite point z_0 on the boundary ∂D of D is called angular (relative to D) if there exists $\varepsilon > 0$ such that every component domain of

$D \cap \{|z - z_0| < \varepsilon\}$ which has z_0 as a boundary point is contained in an angle less than π with vertex at z_0. Angularity at ∞ is similarly defined.

Let $A = A(D)$ be the set of angular points of ∂D relative to D. Obviously A not empty implies that ∂D has positive capacity. The set A(D) can have positive linear measure. E. g. let C be a Cantor set on $|z| = 1$ and let D consist of the open unit disk from which have been deleted all points rz with $z \in C$ and $\frac{1}{2} \leq r < 1$. Then $A(D) = C$ which can, of course, have positive linear measure.

Yet the following holds for arbitrary domains. A(D) is either empty or its harmonic measure, relative to any point of D, is zero. This result follows easily from an unpublished theorem on Brownian paths $\omega(t)$ in the complex plane. This says that almost all such paths have the property that for every real to and every $\varepsilon > 0$ the set of numbers $\frac{\omega(t) - \omega(t_0)}{|\omega(t) - \omega(t_0)|}$ with $t_0 < t < t_0 + \varepsilon$ fills at least an open arc of length π on the unit circle. This theorem is not easy and it would be desirable to give a direct proof of the above result on A(D).

Moreover, the Brownian paths approach will certainly not yield a similar result for the set $B_\alpha(D)$ of ∂D whose points are defined by replacing the angles less than π with translates of $\{x+iy : 0 < x < |y|^\alpha\}$ for a given α with $\frac{1}{2} < \alpha < 1$. (For $\alpha = \frac{1}{2}$ the result is false as can be seen by taking D to be a disk.)

Problem

For which $\alpha \in (\frac{1}{2}, 1)$ is the harmonic measure of the set $B_\alpha(D)$ always zero? (Of course we may assume that the capacity of ∂D is positive.)

A much more difficult problem would be to characterise the monotone functions f(y) having the property that the set obtained on replacing the angles by translates of $\{x+iy : 0 < x < f(|y|)\}$ has necessarily harmonic measure zero.

Similar questions can be asked for Riemann surfaces and for n-dimensional space.

(A. Dvoretzky)

Approximation

7.16 Let γ be a Jordan arc, $d\mu$ a measure on γ. Does the Laplace transform

$$f(z) = \int_{\gamma} e^{z\zeta} d\mu(\zeta)$$

always have 'asymptotically regular' growth as $|z| \to \infty$. The answer might be no. However it could be true anyway that the zeros of $f(z)$ have 'measurable distribution' in the sense of A. Pfluger.

(J. Korevaar)

7.17 Let $f(z)$ be analytic and bounded for $\operatorname{Re} z > 0$; suppose that $|\alpha| < \frac{1}{2}\pi$ and that (r_n) is a sequence of positive integers with $\Sigma \frac{1}{r_n} = \infty$. Show that the exponential type of f on the sequence $z_n = r_n e^{i\alpha}$ is equal to the type of f on the ray $z = re^{i\alpha}$. (Proofs by Boas and by Levinson for $\alpha = 0$ do not seem to work for $|\alpha| > \frac{1}{4}\pi$.)

(J. Korevaar)

7.18 Let Γ be a Jordan curve and suppose that $z = 0$ lies inside it. Wermer showed that when Γ has infinite length, the powers z^n, $n \neq 0$, span all of $C(\Gamma)$. One can indicate conditions on Γ under which the powers z^n, $n \neq n_1, \ldots, n_k$, form a spanning set (S. Korevaar and P. Pfluger). Under what conditions on Γ can one omit an infinite set of powers, and still have a spanning set?

(J. Korevaar)

7.19 For what sets Ω of lattice points (m_k, n_k) do the monomials $x^{m_k} y^{n_k}$ span L^2 or C_0 on the unit square $0 \leq x \leq 1$, $0 \leq y \leq 1$? It is conjectured that the condition $\Sigma \frac{1}{m_k n_k} = \infty$ is sufficient for sets Ω in an angle $\varepsilon x \leq y \leq x/\varepsilon$, $\varepsilon > 0$. The condition is not necessary [Hellerstein].

(J. Korevaar)

Other problems

7.20 (Two constant theorems for the polydisk.) Let $F(z_1, z_2)$ be defined for $|z_1| \leq 1$, $|z_2| \leq 1$ except when $z_1 = z_2$ and $|z_1| = |z_2| = 1$.

Suppose that F is plurisubharmonic in $|z_1| < 1$, $|z_2| < 1$ and that $F(z_1, z_2) \leq \log \frac{1}{|z_1 - z_2|}$, whenever the left hand side is defined. Further suppose that $F(z_1, z_2) \leq 0$ for $\{|z_1| = |z_2| = 1, z_1 \neq z_2\}$. Does it follow that $F(z_1, z_2) \leq 0$ for $|z_1| < 1$, $|z_2| < 1$?

(L. Rubel, A. Shields)

7.21 Suppose that $|z_k| = 1$ $(1 \leq k < \infty)$. Put

$$A_l = \limsup_{m\to\infty} \left| \sum_{k=1}^{m} z_k^l \right| .$$

It is easy to see that there is a sequence z_k for which $A_l < cl$ for all l, and Clunie proved $A_l > cl^{\frac{1}{2}}$ for infinitely many l. Is there a sequence for which $A_l = o(l)$ $(l \to \infty)$?

(P. Erdős)

7.22 Suppose that A and B are disjoint linked Jordan curves in R^3 which lie at a distance 1 from each other. Show that the length of A is at least 2π. The corresponding result with a positive absolute constant instead of 2π is due to the proposer.

(F. W. Gehring)

References

L. V. Ahlfors. [7.11] Extension of quasiconformal mappings from two to three dimensions. Proc. Nat. Acad. Sci., 51 (1964), 768-71.

J. M. Anderson, J. Clunie and Ch. Pommerenke. [5.30] On Bloch functions and normal functions. J. Reine Angew. Math. (1974) (to appear).

A. Baernstein. [6.30] Some extremal problems for univalent functions, harmonic measures and subharmonic functions. C, 6-11.

P. D. Barry. [2] Some theorems related to the cos $\pi\rho$-theorem. Proc. London Math. Soc. (3) 21 (1970), 334-60. [2.36] On a theorem of Besicovitch. Quart. J. of Math. Oxford (2) 14 (1963), 293-302.

J. Becker. [5.31] Löwnersche Differentialgleichung und quasikonform fortsetzbare schlichte Funktionen. J. Reine angew. Math., 255 (1972), 23-43.

A. Besicovitch. [2. 36] On integral functions of order < 1. Math. Ann., 97 (1927), 677-95.

M. Behrens. [5. 27] The corona conjecture for a class of infinitely connected domains. Bull. Amer. Math. Soc., 76 (1970), 387-91.

A. Beurling. [2. 35] Some theorems on boundedness of analytic functions. Duke Math. J., 16 (1949), 355-9. [5. 26] On two problems concerning linear transformations in Hilbert space. Acta Math., 81 (1949), 239-55.

R. P. Boas. [7. 17] Asymptotic properties of functions of exponential type. Duke Math. J., 20 (1953), 433-48.

B. V. Boyarskii. [7. 10] Homeomorphic solutions of Beltrami systems (Russian). D. A. N. S. S. S. R., 102 (1955), 661-4.

R. C. Buck. [2. 43] Integral valued entire functions. Duke Math. J., 15 (1948), 879-91.

L. Carleson. [5. 27] Interpolation by bounded functions and the corona problem. Ann. of Math., 76 (1962), 547-59. [7. 11] The extension problem for quasi-conformal mappings. To appear in Contributions to analysis: a collection of papers dedicated to Lipman Bers, Academic Press 1974.

J. Clunie. [6. 27] On schlicht functions. Ann. of Math. (2) 69 (1959), 511-9. [7. 21] On a problem of Erdős. J. London Math. Soc., 42 (1967), 133-6.

J. Clunie and E. Netanyahu. [4. 22] Personal communication.

Brian Cole. [5. 27] U.

J. Delsarte and J. L. Lions. [2. 45] Moyennes généralisées. Comment. Math. Helv., 33 (1959), 56-69.

P. L. Duren. [5. 23] Theory of H^p spaces. Academic Press N.Y. (1970). [6. 31] Estimation of coefficients of univalent functions by a Tauberian remainder theorem. To be published in J. London Math. Soc.

A. Edrei. [2. 37] A local form of the Phragmén-Lindelöf indicator. Mathematika, 17 (1970), 149-72.

A. Edrei and W. H. J. Fuchs [2. 39] U.

P. Erdős. [2. 46] An interpolation problem associated with the continuum hypothesis. Michigan Math. J., 11 (1964), 9-10.

M. Essén and D. F. Shea. [2.39] Applications of Denjoy integral inequalities to growth problems for subharmonic and meromorphic functions. C, 59-68.

C. Fitzgerald. [5.23] Quadratic inequalities and coefficient estimates for schlicht functions. Arch. Rational Mech. Anal., 46 (1972), 356-68.

O. Frostman. [5.23] Sur les produits de Blaschke. Kungl. Fysiogr. Sällsk. i Lund Förh., 12 no. 15, (1942), 169-82.

T. W. Gamelin. [5.27] Localisation of the corona problem. Pacific J. Math., 34 (1970), 73-81.

F. W. Gehring. [7.12] The L^p-integrability of the partial derivatives of a quasiconformal mapping. Acta Math., 130 (1973), 265-77. [7.22] The Hausdorff measure of sets which link in euclidian space. Contributions to analysis: a collection of papers dedicated to Lipman Bers. Academic Press (1974).

F. W. Gehring and E. Reich. [7.9] Area distortion under quasiconformal mappings. Ann. Acad. Sci. Fenn. Ser. A. no. 388 (1966) 15 pp.

R. L. Hall. [5.34] On the asymptotic behaviour of functions holomorphic in the unit disc. Math. Z., 107 (1968), 357-62.

V. P. Havin. [7.14] Boundary properties of integrals of Cauchy type and of conjugate harmonic functions in regions with rectifiable boundary (Russian). Math. Sb. (N. S.) 68 (110) (1965), 499-517.

W. K. Hayman. [2] The minimum modulus of large integral functions. Proc. London Math. Soc. (3) 2 (1952), 469-512. [2.36] Some examples related to the cos $\pi\rho$-theorem. Macintyre memorial volume (Ohio U. P. 1970), 149-70. [2.38] The minimum modulus of integral functions of order one. To be published in J. d'Analyse. [2.41] Slowly growing integral and subharmonic functions. Comment. Math. Helv., 34 (1960), 75-84. [3.13] Subharmonic functions in R^m. Proc. International Congress of Mathematicians 1970, vol. 2, 601-5. [5.23] On the characteristic of functions meromorphic in the unit disk and of their integrals. Acta Math., 112 (1964), 181-214.

L. I. Hedberg. [5.25] Weighted mean square approximation in plane regions and generators of an algebra of analytic functions. Ark.

f. Math., 5 (1965), 541-52.

S. Hellerstein. [7.19] Some analytic varieties in the polydisc and the Szász problem in several variables. Trans. Amer. Math. Soc., 158 (1971), 282-92.

S. Hellerstein and J. Korevaar. [2.44] The real values of an entire function. Bull. Amer. Math. Soc., 70 (1964), 608-10.

S. Hellerstein and D. F. Shea. [3.13] An extremal problem concerning entire functions with radially distributed zeros. C, 81-8.

B. Kjellberg. [2.38] A theorem on the minimum modulus of entire functions. Math. Scand., 12 (1963), 5-11.

J. Korevaar and P. Pfluger. [7.18] Spanning sets of powers on wild Jordan curves. To appear in Proc. Nederl. Akad. Wetensch.

N. Levinson. [7.17] On the growth of analytic functions. Trans. Amer. Math. Soc., 43 (1938), 240-57.

F. S. Lisin. [5.25] On a question of mean square approximation connected with the study of generators of the algebra l^1 (Russian). Mat. Zametki, 3 (1968), 703-6.

A. J. Lohwater, G. Piranian and W. Rudin. [6.30] The derivative of a schlicht function. Math. Scand., 3 (1955), 103-6.

T. H. Macgregor. [6.30] Applications of extreme-point theory to univalent functions. Michigan Math. J., 19 (1972), 361-76.

D. J. Newman. [5.25] Generators in l_1. Trans. Amer. Math. Soc., 113 (1964), 393-6.

A. Pfluger. [7.16] Über ganze Funktionen ganzer Ordnung. Comm. Math. Helv., 18 (1945/6), 177-203.

G. Piranian. [2.41] An entire function of restricted growth. Comm. Math. Helv., 33 (1959), 322-4. [4.23] U.

Ch. Pommerenke. [4.23] On some problems by Erdős, Herzog and Piranian. Michigan Math. J., 6 (1959), 221-5. [6.27] Lacunary power series and univalent functions. Michigan Math. J., 11 (1964), 219-23.

L. Rubel, A. Shields and B. A. Taylor. [5.28] To be published in the Journal of Approximation Theory.

S. Ruscheweyh and T. B. Sheil-Small. [5.23] Hadamard products of schlicht functions and the Pólya-Schoenberg conjecture.

Comment. Math. Helv., 48 (1973), 119-35.

T. B. Sheil-Small. [6.28] On the convolution of analytic functions. J. reine angew. Math., 258 (1973), 137-52.

L. R. Sons. [5.37] Value distribution and power series with moderate gaps. Michigan Math. J., 13 (1966), 425-33.

F. Sunyer y Balaguer. [5.37] Sur la substitution d'une valeur exceptionelle par une propriété lacunaire. Acta Math., 87 (1952), 17-31.

G. Valiron. [2.39] Sur le minimum de module des fonctions entières d'ordre inférieur à un. Mathematica (Cluj) 11 (1935), 264-9.

G. Weiss and M. Weiss. [5.36] On the Picard property of lacunary power series. Studia Math., 22 (1962-63), 221-45.

A. Weitsman. [1.26] Meromorphic functions with large sums of deficiencies. C, 133-6.

J. Wermer. [7.18] Non-rectifiable simple closed curve. Amer. Math. Monthly, 64 (1957), 372.

J. M. Whittaker. [2.43] Interpolatory function theory. (Cambridge tract no. 33 (1935), C. U. P.)

A. Wiman. [3.11] Über den Zusammenhang zwischen dem Maximalbetrage einer analytischen Funktion und dem grössten Betrage bei gegebenem Argumente der Funktion. Acta Math. 41 (1916), 1-28.